International Conference on Artificial Intelligence and Machine Learning (ICAIML 2020)

Held online due to COVID-19

IOP Conference Series: Materials Science and Engineering
Volume 1049

Jaipur, India
4 - 5 September 2020

ISBN: 978-1-7138-3335-2
ISSN: 1757-8981

Printed from e-media with permission by:

Curran Associates, Inc.
57 Morehouse Lane
Red Hook, NY 12571

Institute of Physics
Dirac House, Temple Back
Bristol BS1 6BE UK

Phone: 44 1 17 929 7481
Fax: 44 1 17 920 0979

techtracking@iop.org

Additional copies of this publication are available from:

Curran Associates, Inc.
57 Morehouse Lane
Red Hook, NY 12571 USA
Phone: 845-758-0400
Fax: 845-758-2633
Email: curran@proceedings.com
Web: www.proceedings.com

TABLE OF CONTENTS

IOP Conf. Series: Materials Science and Engineering 1049 (2021) 011001 doi:10.1088/1757-899X/1049/1/011001

Preface

The International Conference on Artificial Intelligence and Machine Learning (ICAIML 2020) was held at the Jaipur, Rajasthan, India September 04-05, 2020. Due to the evolving public health concerns and travel advisories issued for the avoidance of non-essential travel in view of COVID-19, the conference ICAIML-2020 has been conducted online on virtual platform as participants could not attend the conference in person.

The conference was inaugurated by lighting of lamp and Sarasvati Vandana in virtual presence of Chief Guest Prof. (Dr.) Jemal Hussein Deakin University, Austrailia, Prof. (Dr.) Guandong Xu, Australia, Dr. Celestine Iwendi, Nigeria, Prof. (Dr.) Ghanshyam Singh, Johannesburg, South Africa, Dr. Aditya Singh, Organizing Secretary and Chair, Dr. Pallavi Singh, Conference Chair and Delegates.

Keynote speakers of the conference were:

1. Prof. (Dr.) Guandong Xu, Australia,
2. Prof. (Dr.) Ghanshyam Singh, South Africa,
3. Dr. Nikhil Marriwala, India,
4. Dr. Celestine Iwendi, Nigeria,
5. Dr. Marzieh Khakifirooz, Mexico
6. Prof. (Dr.) Jemal Hussein, Australia

Session Chairs,

1. Dr. Ramesh Kumar, Govt. Engineering College, Ajmer
2. Dr. Anshuman Kalla, Manipal University
3. Dr. Pankaj Goswami, TMU, India
4. Dr. Garima Goswami, TMU, India
5. Dr. Sudhir Sharma, JNU Jaipur
6. Dr. Kalyan Acharyja, JNU Jaipur
7. Dr. Nikhil Marriwala, UIET, KUK
8. Dr. Anshu Gautam, Manav Rachna Univ.

The all tracks of the conference have received 113 papers out of them 68 papers have been shortlisted for presentation. However, only 21 papers have been presented and then selected for publication. There were more than 80 educationists, researchers and students from the country and abroad participated and presented the papers in 4 presentation session tracks during two days. All materials were reviewed by the editors: Dr. Aditya Singh, Professor, Department of ECE, Arya College of Engineering & I.T., Jaipur, Rajasthan, India and Dr. Pallavi Singh, Assistant Professor, School of Allied & Health Sciences, Jaipur National University, Jaipur Rajasthan, India. The conference consisted of plenary presentations for 45 minutes and presentation tracks for 180 minutes. There were 21 paper presentation, 05 oral reports and 05 posters in total. Each presentation has provided extra 5 minutes for questions and answer round. The main objective of the conference is to promote the new findings and research-based outcomes through the techniques, advancement, applications and development of Artificial Intelligence and use of Machine Learning. The series of ICAIML aims to bring together researchers, engineers and practitioners to present the latest achievements and innovations in the interdisciplinary areas of Advanced Computing and

IOP Conf. Series: Materials Science and Engineering 1049 (2021) 011001 doi:10.1088/1757-899X/1049/1/011001

Applied Informatics to discuss thought-provoking developments and challenges, in order to select potential future directions. The major objective of the conference is to encourage the participation of young researchers carrying research in the field relating to soft computing in India and abroad.

We would like to acknowledge all of those who supported ICAIML 2020. The help and contribution of each individual and institution make this conference a great success. In particular, we would like to appreciate the advisory and technical program committee for its valuable inputs and role. We sincerely hope that the ICAIML 2020 proceeding will serve as an important catalyst and research source of references, knowledge and will be useful to readers for their own research work and promoting collaborative research.

December 2020

Aditya Singh

Pallavi Singh

IOP Conf. Series: Materials Science and Engineering 1049 (2021) 011002 doi:10.1088/1757-899X/1049/1/011002

Peer review declaration ICAIML 2020

All papers published in this volume of *IOP Conference Series: Materials Science and Engineering* have been peer reviewed through processes administered by the Editors. Reviews were conducted by expert referees to the professional and scientific standards expected of a proceedings journal published by IOP Publishing.

- **Type of peer review: Single-blind**
  Single-anonymous: reviewers' identities are hidden from authors
  Criteria used: Originality, novelty, motivation, research quality, presentation quality, discussion, theme of the work, Authors had the opportunity to resubmit articles after revision.

- **Conference submission management system:**
  Easy Chair and Direct mailing to conference secretary

- **Number of submissions received:**
  113

- Number of submissions sent for review:
  21 (after presentation and shortlisting)

- **Number of submissions accepted:**
  21

- **Acceptance Rate (Number of Submissions Accepted / Number of Submissions Received X 100):**

  21x100/113= 18.58%

- **Average number of reviews per paper:**
  3

- **Total number of reviewers involved:**
  25

- **Any additional info on review process:**
  Plagiarism software has been used for identification of Plagiarism

- **Contact person for queries:**

  Dr. A. K. Singh
  chair@icaiml.com, icaimlconf@gmail.com

1-D/2-D/3-D Hopfield Associative Memories

[1] **Rama Murthy Garimella,** [2] **Aman Singh,** [3] **Jyothi Prasanna GC,**
[3] **Manasa Jagannadan,** [3]**Vidya Sree Vankam,** [3]**Maha Lakshmi Bairaju**
[1]Mahindra Ecole Centrale, Hyderabad , [2]IIIT-HYDERABAD, Gachibowli, Hyderabad-32,
[3]Rajiv Gandhi University of Knowledge Technologies-RK Valley, Idupulapaya,Andhra Pradesh
E-mail:
[1]rama.murthy@mechyd.ac.in,[2]amanuv19@gmail.com,[3]prasannajyothi805@gmail.com,
[3]manasajagannadan@gmail.com, [3]vankamvidyasree@gmail.com, [3]lakshmiiiit49@gmail.com

Abstract. In this research paper storage as well as retrieval of 1-D/2-D/3-D information using Hopfield type Associative Memories (AMs) is discussed. Various Artificial Neural Networks(ANN) architectures are proposed. Also, implementational issues associated with those Associative Memories are discussed. Cascade connection of AM and Convolutional Neural Network is proposed for noise immunity.
Keywords: Associative Memories, Hopfield Network, Ceiling Neuron

1. Introduction

In an effort to the model, biological memory, Hopfield proposed an Associative Memory (the so-called Discrete Time Hopfield Neural Network(DTHNN)). The proposed Associative Memory (AM) is based on a vector of {+1,-1} as the state of the dynamical system. Thus, in such an Artificial Neural Network(ANN), only one dimensional information can be stored.

In [8], the author proposed the design of Multi-Dimensional Neural Networks. A natural question that arises is how to design an associative memory which can store two/three dimensional information. As a first attempt to answer the question, researchers attempted designing associative memory with {+1,-1} arrays as the state of the system. Further some researchers studied associative memories with multi-state neurons, whose state vector has more than two elements [11], [9]. These efforts have many applications for content based Image Retrieval and other problems.

In this research paper we propose storage/retrieval of 1-D/2-D/3-D information from 1-D/ 2-D/ 3-D queries based on the ideas of associative memory.

This research paper is organized as follows. In Section-II, review of related research literature is presented. In Section-III, an architecture based on Parallel Hopfield Neural Network is proposed. In Section-IV, we proposed new architectures by stacking of parallel Hopfield Associative Memories for conversion of lower dimension to higher dimensions. We also discussed on Ceiling Neuron based Associative Memory. In Section-V, a novel deep learning architecture is proposed. In Section-VI, applications are discussed. In Section-VII, results of all the proposed architectures are discussed and implemented on Black & White Images. The research paper concludes in Section-VIII.

2. Review of Related Literature

In research literature on Associated Memories, many researchers realized that Hopfield Associated Memory is very restricted from the point of view of state space. Aizenberg et al proposed multi-state neuron based associative memories. In most of these efforts, the state of the neural network (dynamical System) was a vector. In [8], the author proposed the design of multi-dimensional associative memories. In fact convergence theorem of multi-dimensional Hopfield neural network was proved [3]. 2-D as well as 3-D associative memories find many applications to store images as well as video data [2]. In this research paper, we attempt simple models of 2-D or 3-D associative memories in the spirit of Hopfield's effort.

3. Parallel Hopfield Neural Network

In this section, we propose a variation of DTHNN. The architecture proposed in this section is motivated by the concept of CEILING NEURON proposed in [4]. In this model, there are multiple thresholds at each neuron instead of a single one. We utilize such an idea to arrive at the following nonlinear dynamical system which acts as a two dimensional associative memory.

3.1. Architecture-1 :

$$\tilde{V}(n+1) = Sign\{\bar{W}\tilde{V}(n) - \tilde{T}\} \quad (1)$$

with, $\tilde{V}(0)$ as the initial state matrix. In (1), $\{\tilde{V}(n)$:for $n \geq 0\}$ is a {+1,-1} valued state matrix and $\tilde{T}$ is a matrix of thresholds (motivated by the idea in [4]). It should be noted that (1) corresponds to fully parallel mode of operation of 2-D AM. It readily follows that serial mode of operation of such an associative memory corresponds to updating just one component of $\tilde{V}(n+1)$. We ensure that the diagonal elements of weight matrix, W are all non-negative.

Based on the convergence theorem of ordinary HNN in the serial mode of operation, 2-D AM converges to a stable state (matrix of +1,-1) and to a cycle of length 2 in the fully parallel mode of operation. A more general model of AM motivated by the idea in [6] is the following one:

$$\tilde{V}(n+1) = Sign\{\bar{(}W(n)\tilde{V}(n) - \tilde{T}(n)\}, \quad (2)$$

In the spirit of Parallel HAM in (1), we can use three/higher-dimensional state tensors. Thus, we have the following general multi-dimensional associative memory.

$$\tilde{V}(n+1) = Sign\{\bar{W} \circledast \tilde{V}(n) - \tilde{T}\}, \quad (3)$$

where$\{\tilde{V}(n)$: for $n \geq 0\}$ are state tensors and '$\circledast$' denotes suitable inner product. Also $\tilde{T}$ is the tensor of thresholds.

<u>**Note:**</u> Synthesis of Hopfield Associative memory with desired stable states, when threshold vector is a zero vector was documented in [1] and when the threshold vector is a non-zero vector was documented in [10] naturally generalises to tensor based linear operators with eigen values and eigen tensors. These results naturally apply for architecture-1 above.

4. Stacking of Parallel Hopfield Associative Memories

In the two-dimensional associative memory proposed in the above section, the synaptic weight matrix remains same for updating the state vectors in parallel. An effort to relax this assumption leads to the following architecture.

4.1. Architecture -2

We propose stacking based architecture where the input of 'M' HAMs/HNNs is a matrix of {+1's,-1's} i.e., Let $\tilde{V}(0)$ be a {+1,-1} component matrix. $\tilde{V}(0) = [V_1(0) \quad V_2(0) \ldots\ldots\ldots \quad V_M(0)]$
<u>**Note:**</u> In this architecture, a two dimensional {+1,-1} matrix is associated with a 2-D matrix of {+1,-1}s whose columns correspond to the stable states of associative memories.

4.2. Architecture-3

In this architecture of stack, at each level of stack, we have a different 2-Dimensional associative memory with different initial state matrix that converges to a different stable state matrix. **Note:** 3-Dimensional initial state tensor is associated with a 3-Dimensional stable state tensor.

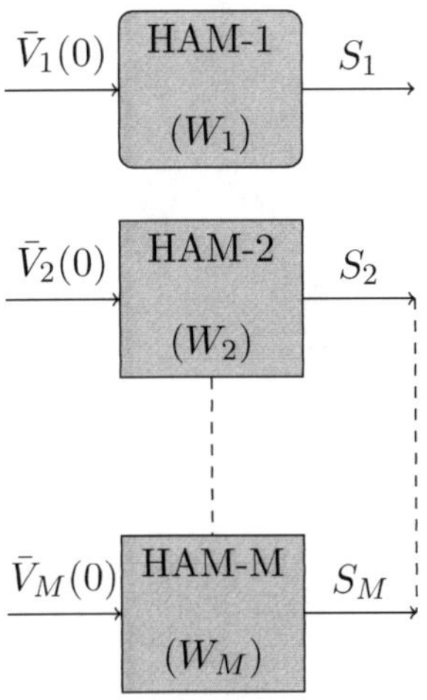

Figure 1: Block diagram of Architecture-2

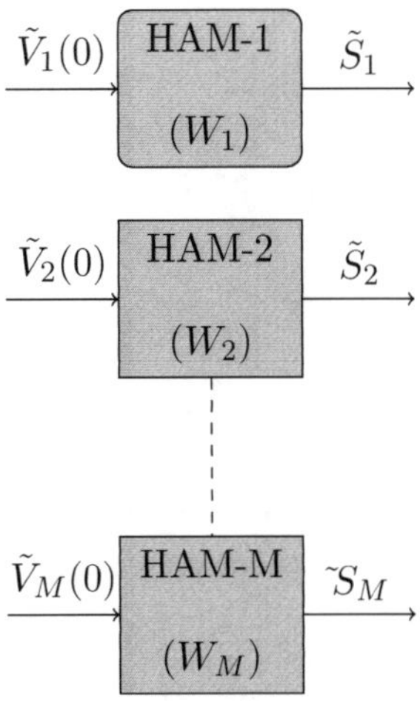

Figure 2: Block diagram of Architecture-3

It is possible to conceive architectures in which a higher dimensional information is associated with a lower dimensional information. We have following three cases:

- 2-Dimensional to 1-Dimensional
- 3-Dimensional to 2-Dimensional
- 3-Dimensional to 1-Dimensional

Such associative memories require higher dimensional stable states to be reduced to lower dimensional stable states.

CEILING NEURON - BASED ASSOCIATIVE MEMORY : 2-D to 1-D

Associative Memory

In Ceiling neuron model [4], every neuron has multiple thresholds. Using multiple thresholds, the net contribution (i.e., $\sum_{i=1}^{N} W_i x_i$) is thresholded and the resulting state of the network is a matrix. In Section-5 of [4], we propose an AM where the state is a vector. Thus, in such an AM, 2-D state information is associated with a 1-D stable states. We, thus expect design of AMs where higher dimensional information is associated with lower dimensional information.
Note-1: When functions of human memory are understood, it becomes evident that higher dimensional information is "associated" with lower dimensional information (and vice-versa) in an effortless manner. One of our goals in this research paper is to arrive at models of ANN which can achieve these functions.

Note-2: Using Parallel, Stacked HNN architectures, 1-D/2-D/3-D information can be stored and retrieved. Thus, a total of Nine architectures[3 x 3] are possible and included three of them.

Note-3: With two stages of Stacked/ parallel architectures, there are twenty seven (9 X 3) possible associative memories. The effort is to model, biological associative memories.

5. Novel Associative Memories : Deep Learning

In most of the applications 1-D/2-D/3-D data (vectors/matrices/3-D arrays) is corrupted by noise [5]. In the case of 1-D neural networks, the author proposed the concept of "HYBRID Neural Networks" [7]. In that research paper, AM (e.g. HNN) is utilized to filter the noise. The input vectors after filtering are fed to a Multi-Layer Perceptron, which performs classification. Generalizing the idea, we employ 2-D/3-D associative memory to filter noise from images, videos. The filtered input is fed to a Deep Convolutional network for performing classification.

The Block diagram representation of such a Deep neural network is provided below.

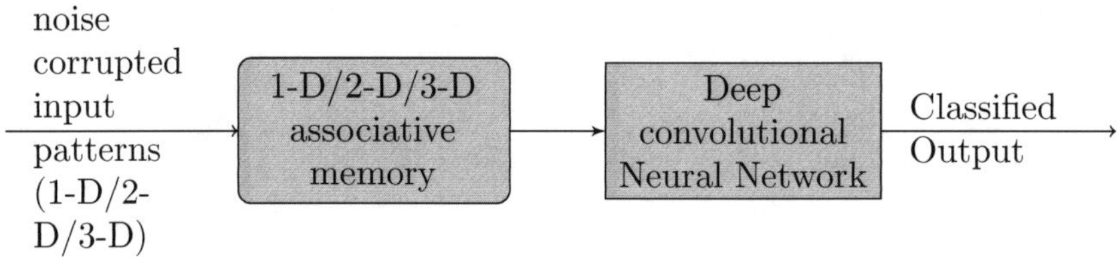

Figure 3: Block diagram of novel associative memories.

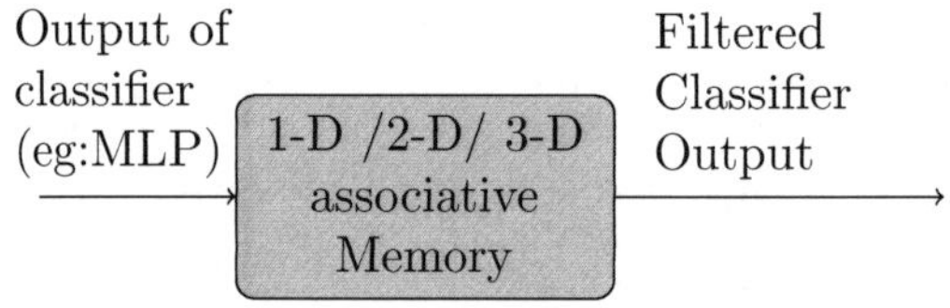

Figure 4: Block diagram of Filtered Classifier Output.

The associative memories discussed above, reach stable state given an input. In the fig(3), the stable states constitute the input CNN. Specifically, the associative memory can be synthesized with 'N' desired stable states each belonging to a class in the classification problem. The CNN can be trained to perform the desired classification problem. We have experimented with the architecture in Fig.4 and achieved good classification performance. The spurious stable states are eliminated by properly training the CNN architecture.

Note: The stable states can be one/two/three dimensional corresponding to audio/text, image, video related data. The synthesis approach proposed in [10] for programming desired stable states can easily be generalized to 2-D/ 3-D associative memories.

In [7], the author also proposed an architecture in which the noisy output patterns of a Multi-Layer Perceptron (MLP) are filtered using an associative memory.

6. Applications

It is well known that the DTHNN (based on convergence theorem) was successfully utilized to store 1-D patterns (which are specified using (+1's, -1's). Generalizing the idea, we innovated the three architectures specified above. These architectures are an effort to model the memory mechanisms in homo-Sapiens. For instance, given the speech input of a person, the memory can correlated with the face image of the person i.e., the biological neural network stores the correspondence between speech signal and the associated image.

In general, human brain is capable of retrieving 1-D/ 2-D/ 3-D information, given the 1-D/

2-D/ 3-D information/signal as the input through the process of association. Hence, out of Nine possible architectures, we provided three architectures in the above sections.

Further, as we can expect, human brains can be expected to have various associative memory units connected to one another using certain "Network Topology ". Such, interacting associative memories are potentially capable of storing heterogeneous types (speech/image/video signals) of information.

7. Results of Implementation

We are currently investigating dedicated purpose hardware to implement 1-D /2-D /3-D associative memories, which can potentially "Speed up" retrieval of stored 1-D/ 2-D/3-D information. We now provide some numerical results on data. We also provide some results related to implementation based on black and white images. The results and implementation of all the architectures mentioned in Section-III and Section-IV are provided below:

7.1. Architecture-1

We have taken 3x3 symmetric weight matrix, one state matrix having elements of +1,-1's and one threshold matrix whose values are in the range of 0 to 1. We implemented AM in parallel mode of operation. Here we have taken, $W = \begin{bmatrix} 0 & -3 & -2 \\ -3 & 0 & -4 \\ -2 & -4 & 0 \end{bmatrix}$. As a result, it converges to a cycle of length 2.

7.2. Architecture-2

We have taken two HAMs having same or different symmetric 3x3 weight matrices and a state matrix of two different 3x3 state vectors are $\tilde{V}(0) = [\bar{V}_1(0) \quad \bar{V}_2(0)]$. We have taken,

$W_1 = \begin{bmatrix} 0 & 1 & 2 \\ 1 & 0 & 3 \\ 2 & 3 & 0 \end{bmatrix}$ and $W_2 = \begin{bmatrix} 0 & 4 & 5 \\ 4 & 0 & 6 \\ 5 & 6 & 0 \end{bmatrix}$. Such an AM converged with cycle of length 2 in parallel mode of operation. Then we stack all the final state vectors, which leads to a matrix of 2-D.

7.3. Architecture-3

We have taken two HAMs having same or different symmetric 3x3 weight matrices and a state matrix of two different 3x3 state vectors as $\tilde{V}_1(0)$ and $\tilde{V}_2(0)$.

We have taken, $W_1 = \begin{bmatrix} 0 & 4 & 5 \\ 4 & 0 & 6 \\ 5 & 6 & 0 \end{bmatrix}$ and $W_2 = \begin{bmatrix} 0 & 1 & 2 \\ 1 & 0 & 3 \\ 2 & 3 & 0 \end{bmatrix}$.Such AM converged with cycle of length 2 in parallel mode of operation. Then we stack all the final state matrices, which leads to 3-D. Finally it is shown that 3-D input converges to 3-D output.

7.4. Implementation on Black and White Images:

Here we have taken a black & white image having size of 5x5 . Then we converted it into a 5x5 matrix of elements {+1,-1}. Then it is given as input to HAM. It converges in parallel mode with cycle of length 2. Then we fix the weights.

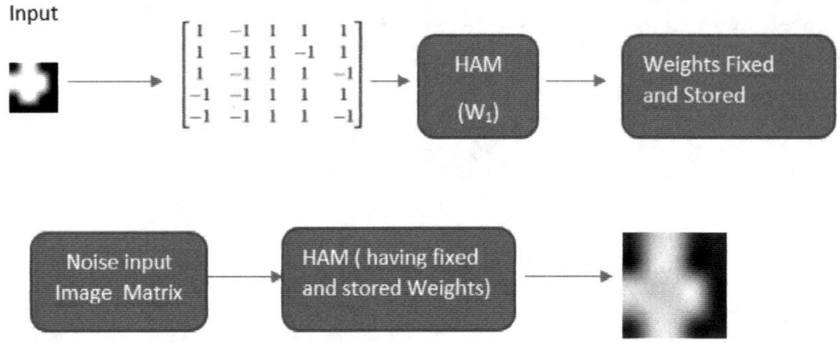

Figure 5: Block diagram of implementation on Black and White image

Now, add some noise to the input picture. Then apply this to the HAM having fixed weights. Finally, the original picture is reproduced. It is explained briefly in Fig.4.

8. Conclusions

In this research paper, the concepts of parallel, stacked associative memories are discussed. Various novel architectures of Associative Memories for storage/retrieval of 1-D/2-D/3-D information are discussed. Artificial Neural Network architecture based on cascading of AM's, CNN's are proposed. It is expected that these architectures will be of practical utility. We are actively investigating on remaining six architectures.

References

[1] Garimella Rama Murthy , Moncef Gabbouj , "On the design of Hopfield Neural Networks: Synthesis of hopfield type associative memories," Proceedings of IEEE International Joint Conference on Neural Networks (IJCNN 2015), July 2015 , Appears on IEEE Explore.

[2] H. Sun , H. Hasegawa , I. Yamada , "Multidimensional associative memory neural network to recall nearest pattern from input [image matching example]," 2005.

[3] G. Rama Murthy , "Multi / Infinite Dimensional Neural Networks, Multi/ Infinte Dimensional Logic Theory," Published in 1995 IEEE International Symposium on Circuits and Systems, Seattle.International Journal of Neural Systems,Vol.14.No.3,pp.223-225,2005...Generalization of Boolean Logic Theory.

[4] G.Rama Murthy , M. Dileep ,R.Anil , "Novel Ceiling Neuron Model and its Application," proceeding of IEEE International Joint Conference on Neural Networks (IJCNN),2019,July 2019, Budapest,Hungary.

[5] G.Rama Murthy , M. Dileep ,R.Anil , "Convolutional Associative Memory," International Conference on Neural Information Processing [ICONIP 2015], November 2015, Turkey.....

[6] G.Rama Murthy , L.Behra , "Adaptive Associative Memory," Published in GESTS International Transactions on Communications and Signal Processing, April 2006. Also, in Proceedings of National Conference on Soft Computing, Bhubaneswar, 24th- 26th March-2006.

[7] G.Rama Murthy , "Hybrid Neural Networks," Proceedings of International Conference on Power System Analysis, Control and Optimization (PSACO-2008), 13th- 15th March-2008.

[8] G.Rama Murthy , "Multi-Dimensional Neural Networks: Unified Theory," New Age International (P) Limited, Publishers.Publishedin November 2007.

[9] Igor Aizenberg ,"Complex-Valued Neural Networks with Multi-Valued Neurons," Proceedings of International Conference on Power System Analysis,2011.

[10] Rama Murthy Garimella , Krishna Vamshi Reddy Lankala , Devaki Nimmagadda , Sree Divya Bondalapati , "Synthesis/Programming of Hopfield Associative Memory," ICMLDS-2019 , ACM Digital Library , December 2019.

[11] Mehmet Kerem Müezzinoglu , Cüneyt Güzelis , Jacek M. Zurada ,"A New Design Method for the Complex-Valued Multistate Hopfield Associative Memory," IEEE TRANSACTIONS ON NEURAL NETWORKS, VOL. 14, NO. 4, JULY 2003.

Industrial Temperature Measurement of Flame and Flue gases using IOT and Image processing

K. Sujatha[1a], R.S. Ponmagal[1b], K. SenthilKumar[1c], V. Karthikeyan[1d], K. S. Ramkumar[2a], N P GBhavani[2b]

[1]Professor, Dept. of EEE., Dr. MGR Educational and Research Institute, Tamilnadu.
[1]Associate Professor, Dept. of CSE., SRM Institute of Science and Technology
[1]Professor, Dept.of ECE., Dr.MGR Educational and Research Institute, Tamilnadu.
[1]Professor, Dept.of EEE., Dr.MGR Educational and Research Institute, Tamilnadu.
[2]Dept.of Electronics and Communication, J.N.N Institute of Engineering, Tamilnadu.
[2]Associate Professor, Dept.of EEE, Meenakshi College of Engineering, Tamil Nadu

drksujatha23@gmail.com

Abstract. This study effort contracts with measuring the flame temperature in the combustion chamber along with a scheme to monitor the temperature of flue gases from the power plants. Presently, the gases from the exhaust are monitore dusing gas analyzers which are positioned at the exit point of the chimney. These gas analyzers are prone to cold end corrosion which will lead to malfunctioning of the equipment causing erroneous results during measurements. Thiskind of indigenous technique facilitates to develop an Industrial IoT (IIoT) temperature monitoring system at power plants. A combination of Principal Component Analysis (PCA) with Cascaded Neural Network (CNN) is used. The features extracted and reduced serve as inputs to CNN. An efficient technology to scrutinize and measure the temperature inside the combustion chamber thereby reducing the amount of flue gas emissions by the partnership of two methods.

Keywords:*Temperature measurement, Flue gases, Power plants, Cascaded Neural Network, Service Oriented Architecture, Internet of things and Principal component Analysis.*

1. OVERVIEW

This investigationis about monitoring and measuring of flame temperature and exhaustusing indigenous algorithms. The flame in furnace is generally observed using an Infrared camera placed tangentially at various levels of the boiler to monitor the existence of the flame during the start-up lighting process of the boiler [2]. Exhaust gases from the industrial boilers can cause air pollution and also depletion of ozone layer[9, 11].

2. MATERIALS AND METHODS

The values of the correlation coefficients corresponding to complete and incomplete combustion categories of flame images are recorded in Table 1.It is a dimensionality reduction method that is often used to reduce the dimensionality for wide range of data sets, by converting these variables into a small values with useful information. If the number of variables are reduced then the accuracy need to be compromised, but this dimensionality reduction facilitates a simple procedure by sacrificing a small amount of accuracy [7]. If data set has a smaller dimension, it is simple to analyze and interpret the outccomes at a faster rate using machine learning algorithms [4,5,6].

CNN comprises of a network with nodes in concealed layer which are included one at a time and are not altered after their inclusion. The word "cascade" indicates that the output from all nodes feed the newly created nodes. A CNN is said to have three layers namely the contribution Layer has a vector of values ($x_1 \ldots x_p$) called as predictor variables. The nodes in contribution layer performs no action and functions as a distributor to the nodes in the concealed and production layers. Constant inputs like 1

IOP Conf. Series: Materials Science and Engineering 1049 (2021) 012002 doi:10.1088/1757-899X/1049/1/012002

and 0, are used as bias which serves as inputs to the concealed and production layer nodes. The weights which are the random number between '0' and '1' are used to multiply the bias and add the sum to value entering the node. The nodes entering the concealed layer have a quantity from each input neuron, multiplied by a weight, and the resulting weighted values are summed up together to obtain the net value. This net value is passed over the thrashing function which outputs a value. The outputs from the concealed layer are distributed to the production layer. In case of regression the production Layer has only one neuron whereas for classification based applications, for each type there is a node which represents a target value [13].

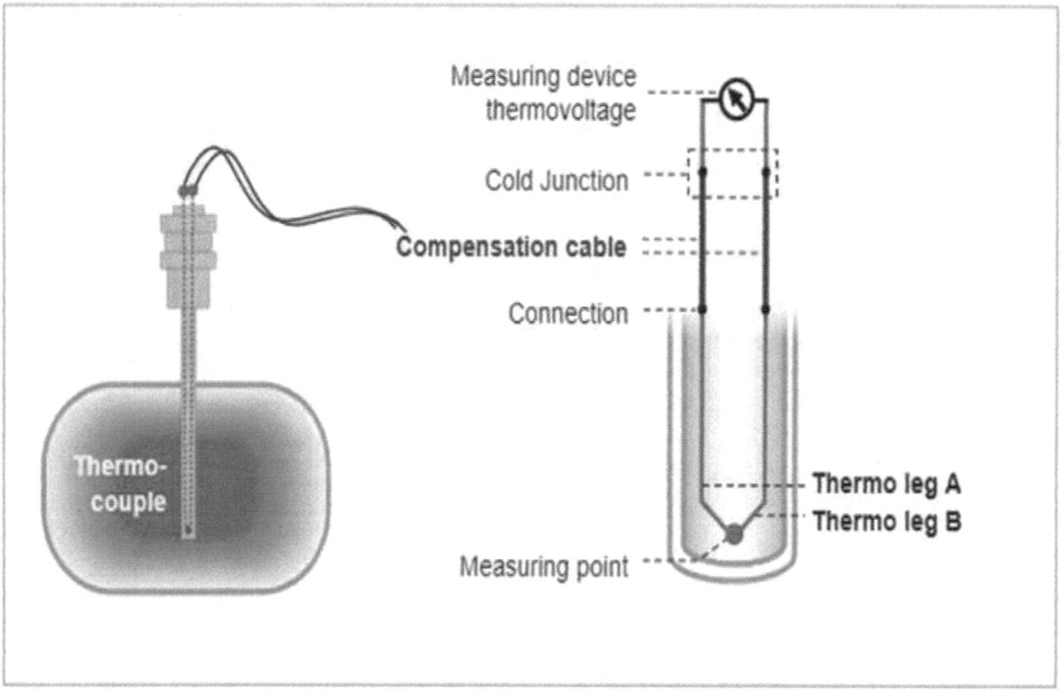

Figure 1.Existing system for Temperature measurement in Power plants

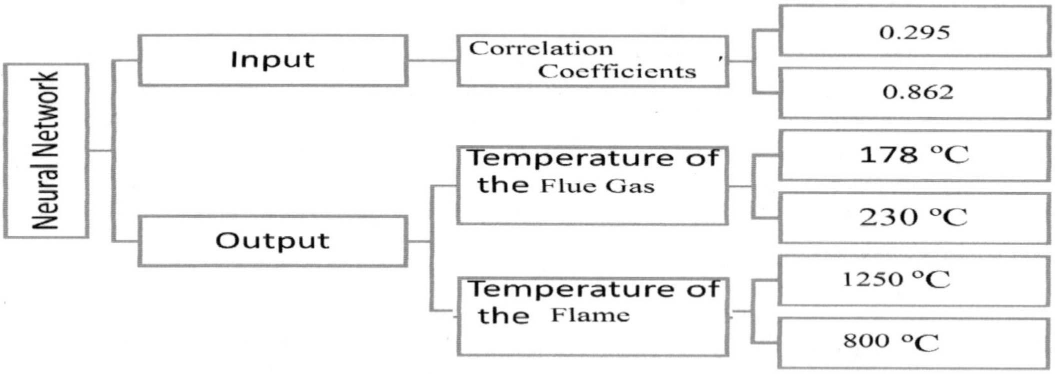

Figure 2.Diagramatic demonstration of flame scrutinizing method

Table 1 Flame and Flue gas emissions Temperature

S. No	Images of flame	Flame Temperature in ^{0}C	Flue gas Temperature in ^{0}C	Flame image Category
1.		1250	230	Complete combustion
2.	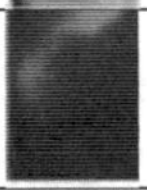	350	121	Incomplete combustion

3. RESULTS AND DISCUSSION

The performance of Euclidean Distance Classifier is shown in Figure 3. The CNN consists of only one node in the contribution and production layer at the beginning. There are no nodes in the concealed layer. Every node in the contribution layer is connected to every node in production layer by a connection whose coefficient value is adjustable [11].The classification is done with the feature set ranking 1,2,3,4 and 1,2,3,4,5,6. Table 2 shows the estimation constraints for classifiers.

Table 2. Estimation constraints for classifiers

S.No	**Parameters**	**CNN**	**Euclidean classifier**
1	Sample size	51	51
2	No. of images classified correctly (%)	100	90
3	Kappa value	1	0.8485
4	signify total error	0.0359	0.077
5	Root signify squared error	0.0391	0.1946
6	Relative total error	0.0826	0.1737
7	Root relative squared error	0.08233	0.4177

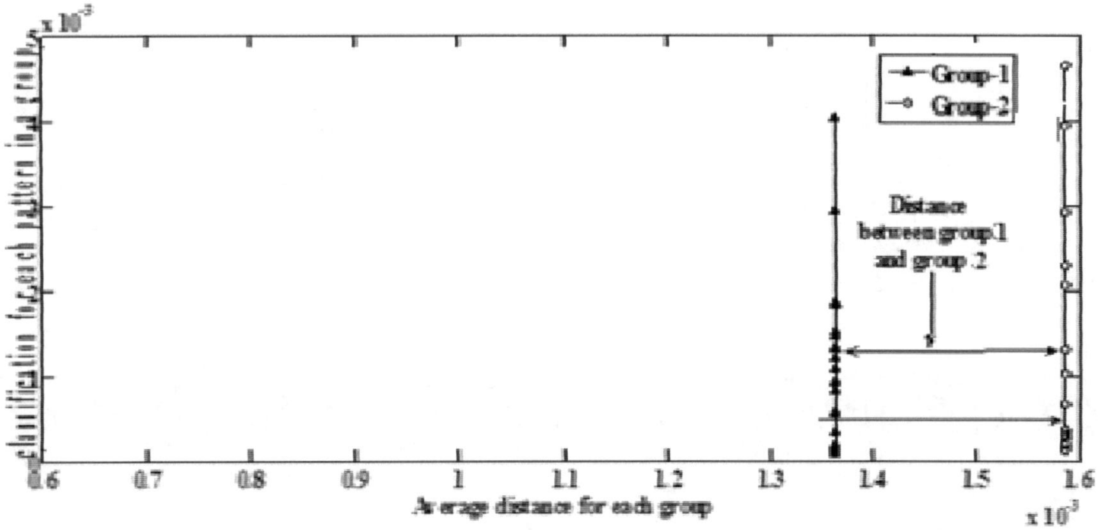

Figure 3. Patterns allocation for all the three classes using Euclidean classifier

IOP Conf. Series: Materials Science and Engineering 1049 (2021) 012002 doi:10.1088/1757-899X/1049/1/012002

Training and testing the CNN classifier and Euclidean classifier is indicated in Table 2. Testing the Euclidean classifier is done with 27 images. Results for testing the Euclidean classifier is shown in figure 3 with the feature set 1, 2, 3 and 4 and similarly the testing of CNN is also done. The comparsion results for testing the Euclidean and CNN classifier proves that uncertainty matrix for the Euclidean. The performance of the CNN classifier as represented in Figure 5.

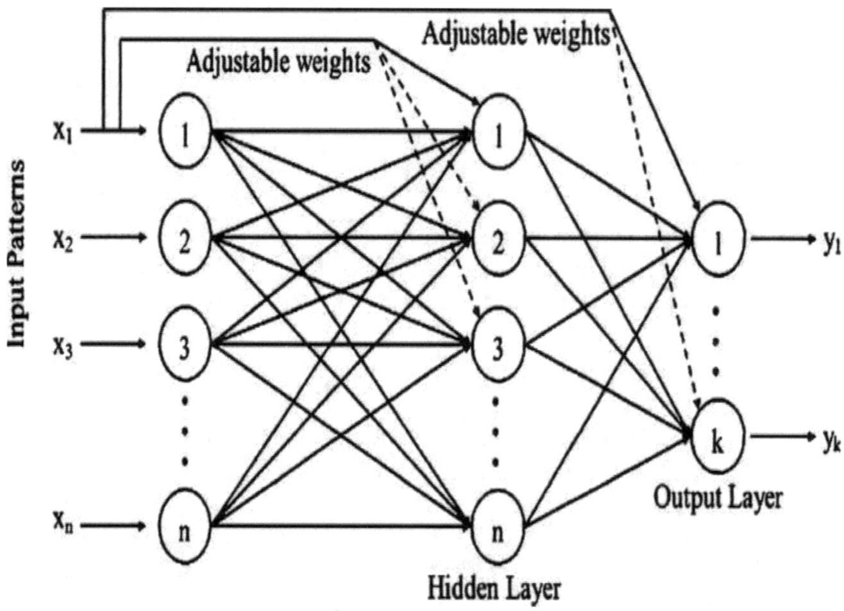

Figure 4. Network Architecture for CNN

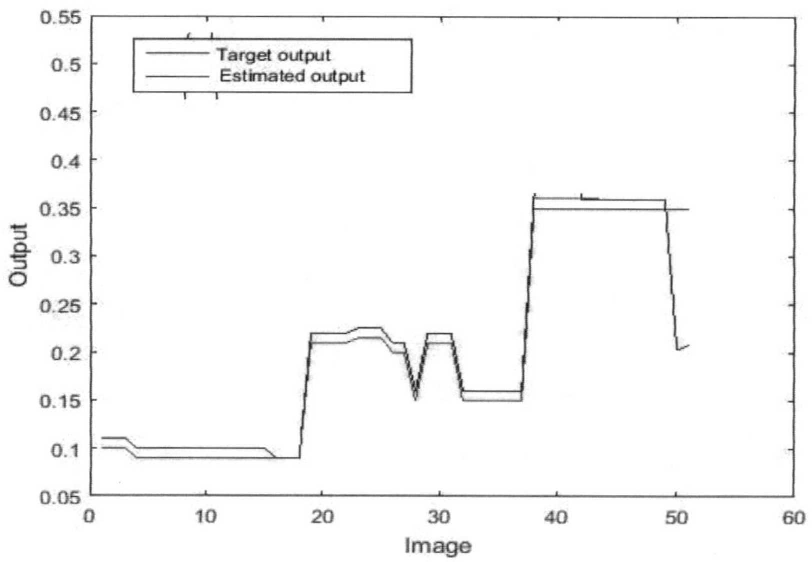

Figure 5. Output for CNN

Table 3. Estimation constraints for classifiers 1,2,3,4,5,6

S.No	constraints	CNN	Euclidean classifier
1	Total no.of instances	51	46
2	Instances classified correctly (%)	100	90.16
3	Kappa value	1	0.8496
4	signify total error	0.029	0.0785
5	Root signify squared error	0.0377	0.2183
6	Relative total error	0.067	0.1807
7	Root relative squared error	0.0877	0.4688

4. CONCLUSION

The flame images for a boiler in the power station were obtained from the control room via the Service Oriented Architeure.Using PCA the extracted features are reduced. With 51 images taken from Complete and incomplete combustion the training is done . CNN classifier is inferred to yield maximum classification performance

References

[1] Hong Z.Q.and Yang Y.J., "Optimal Discriminate Plane for a Small Number of Samples and Design Method of Classifier on the Plane," pattern recognition, vol. 24, pp. 317 – 324, (1991).
[2] Foley D.H., "Consideration of Sample and Feature Size," IEEE Trans., on Info. Theory, vol.18, no.5, pp. 626 – 681, (1972).
[3] Lippmann R.P., An introduction to computing with neural nets, IEEE Trans. on Acoustics, Speech and Signal Processing Magazine, vol35, no 4, pp.4.-22, (1987).
[4] Purushothaman S. and Srinivasa Y.G., A procedure for training artificial neural network with the application of tool wear monitoring, Int. J. of Production Research, vol.36, no.3, pp.635-651, (1998).
[5] Meng Joo Er,.Shiqian Wu, Juwei Lu and Hock Lye Toh, Face Recognition with Radial Basis Function (RBF) Neural Networks, IEEE Trans. on Neural Networks, vol. 13, no.3,pp. 697 - 910, (2002).
[6] Pu Han, Xin Zhang, Chenggang Zhen and Bing Wang, Boiler Flame Image Classification Based on Hidden Markov Model, IEEE ISIE, (2006).
[7] David Meyer, Friedrich Leisch, and Kurt Hornik, The support vector machine under test, Neurocomputing 55(1-2), 169-186, (2003).
[8] Lu G, Gilabert G and Yan Y, Vision based monitoring and characterization of combustion flames, Journal of Physics: Conference Series 15, pp. 194–200, (2005).
[9] WÓJCIK.W, Application of fibre-optic flame monitoring systems to diagnostics of combustion process in power boilers, Bulletin of The Polish Academy Of Sciences , Technical Sciences, Vol. 56, No. 2, (2008).
[10] Rish, Irina. (2001). An empirical study of the naive Bayes classifier, IJCAI Workshop on Empirical Methods in Artificial Intelligence, (2001).
[11] Y. Guo, R. Greiner, Discriminative Model Selection for Belief Net Structures, AAAI, pp. 770-776, (2005).

IOP Conf. Series: Materials Science and Engineering 1049 (2021) 012003 doi:10.1088/1757-899X/1049/1/012003

Bitcoin Financial Forecasting

Dhyanendra Jain[1], Ashu Jain[2], Amit Kumar Pandey[3], Jogender Kumar[4]
Dr. Akhilesh Das Gupta Institute of Technology & Management, New Delhi

[1]dhyanendra.jain@gmail.com, [2]ajainashu@gmail.com, 3amitpandey33@gmail.com, 4jogender@rediffmail.com

Abstract. Bitcoin is one of the crypto currencies and is most unpredictable currencies. In the world of crypto currencies, the value of a coin can unpredictably upgrade or degrade. In this study, the model has been trained to predict the value of Bitcoin in USD at any given time stamp. For this prediction, three algorithms of machine learning - Linear Regression (LR), Support Vector Regression (SVR) and Neural Network Regression (NNR) have been used. The model is trained using the collected dataset. After the collection of data set, we first applied SVR algorithm, then we used LR and then NNR to calculate the error compared to the actual value. Root mean squared error (RMSE) is used as the predictive measure. Out of the three algorithms, LR was found out to be more accurate for predicting the value of bitcoin.

1. Introduction

Bitcoin is an inclining crypto currency on the planet with appropriation developing reliably after some time. Because of the open idea of bitcoin it likewise represents another worldview instead of customary financial markets. It works on a decentralized, shared and trustless framework in which all transactions are presented on an open record called the Blockchain. This kind of straightforwardness is incredible in other financial markets. Foreseeing the time series is the use of a model to predict the upcoming values which are dependent on the current values. In the crypto currency world, costs are exceptionally unstable. This implies the estimation of a coin can go up or down actually rapidly, with regularly no clarification concerning why.

Prediction of develop financial markets, for example, the securities exchange has been looked into finally. Bit coin displays an intriguing corresponding to this as it is a period arrangement expectation issue in a market still in its transient stage. Subsequently, there is high instability in the market and this gives an open door as far as expectation. Customary time arrangement expectation techniques depend on linear suppositions and require information that can be separated into pattern, occasional and clamor to be compelling. This kind of procedure is increasingly appropriate for an errand, for example, anticipating deals where regular impacts are available. Because of the absence of regularity in the Bit-coin market and its high instability, these techniques are not successful for this undertaking. Given the multifaceted nature of the assignment, Machine learning (ML) makes for a fascinating innovative arrangement dependent on its exhibition in comparable zones. Assignments, for example, natural language processing which is additionally successive in nature and has demonstrated promising outcomes. This kind of assignment utilizes information of a consecutive sort and thus is like a value expectation task. All through this undertaking, Linear Regression (LR), Neural Network Regression (NNR) and Support Vector Regression (SVR) have been utilized to show up at the ideal aftereffect of anticipating the bitcoin.

Through this paper we are aiming to train our model to predict the value of the bitcoin at any given instant of time and see the error value as compared to the actual value at that instant. For this purpose, root mean squared value of the error has been calculated.

Organization of the paper is : Section 2 includes the literature survey done for the predictive analysis using machine learning algorithms. Section 3 gives the overview of the models used. Section 4 discusses about the dataset used. Section 5 includes the predictive analysis. Section 6 includes conclusion and directions of future work.

2. Related Work

Günay and Ensarı [1] broke down the notable ML calculations which are for the most part utilized in the past investigations to plan another model to foresee customer churn. Kanawaday and Sane[2] investigated the utilization of AutoRegressive Integrated Moving Average (ARIMA) anticipating on the time arrangement information gathered from different sensors from a Slitting Machine, to foresee the potential disappointments and quality deformities, hence improving the general assembling process. Charleonnan et al [3] presented ML strategies for anticipating the chronic kidney disease utilizing clinical information. Four ML techniques are investigated which includes K-closest neighbors (KNN), support vector machine (SVM), logistic regression (LR), and choice tree classifiers. These models are developed by using the dataset of chronic kidney disease and then all the models are compared to select the best one. Rejawat et al [4] proposed a hybrid ML procedure to classify individuals into patient and healthy classes, separately. The patient class is the people having diabetes and the healthy class is the people having no diabetes. Proposed model was assessed using various implementation measures. Pushpa et al [5] utilized the idea of ML to anticipate the consequence of class students. In view of the presentation of the students in past semester, and the scores of internal examinations of the current semester, the final outcome, regardless of whether the understudy passes or fails the current semester is registered before the final assessment really happens. Hassan et al [6] showed up with the point of aggregating a major piece of the information about solar power anticipating, concentrating on the latest headways and future patterns. Karasu et al [7] performed predictive analysis of bitcoin value with LR and SVM from ML techniques by utilizing time arrangement comprising of every day Bitcoin shutting costs during 2012-2018. Velankar et al [8] endeavored to anticipate the Bitcoin cost precisely thinking about different parameters that influence the Bitcoin value. For the principal period of our examination, they expected to comprehend and distinguish day by day inclines in the Bitcoin showcase while picking up understanding into ideal highlights encompassing Bitcoin cost. Sakiz and Kutlugun [9] in the wake of clarifying the fundamental ideas driving distributed architetcure and blockchain innovation behind crypto cash, artificial intelligence calculations were exploited, and dependent on most recent three years values of bitcoin determining was performed for Bitcoin which has a colossal piece of the overall industry in since nine years.

3. Models Used

3.1. Linear Regression

Regression is a strategy for exhibiting a target esteem dependent on independent pointers. This strategy is generally used for assessing and finding conditions and legitimate outcomes association between factors. Regression methods generally differentiate dependent on the amount of independent components and the sort of association between the independent and dependent variables.

Direct linear regression is a kind of regression assessment where the amount of independent components is one and there is a linear association between the independent(x) and dependent(y) variable.

3.2. Neural Network Regression

Albeit neural systems are comprehensively known for use in profound learning and showing complex issues, for instance, picture acknowledgment, they are viably changed in accordance with regression issues. Any class of quantifiable models can be named a neural system in case they use adaptable loads and can assessed non-linear components of their data sources. Along these lines neural system regression is fit to issues where an inexorably standard regression model can't fit an answer.

Neural system regression is a directed learning procedure, and subsequently requires a named dataset, which consolidates a name area. Since a regression model predicts a numerical worth, the name fragment must be a numerical data type.

You can prepare the model by giving the model and the labeled dataset as a contribution to Train Model or Tune Model hyper parameters. The prepared model would then be able to be utilized to foresee values for the new information models.

Neural network wording is enlivened by the natural activities of specific cells called neurons. A neuron is a cell that has a few information sources that can be enacted by some outside procedure. Contingent upon the measure of activation, the neuron creates its own action and sends this along its yields. What's more, explicit info or yield ways might be "fortified" or weighted higher than different ways. The speculation is that since the human mind is only a network of neurons, we can imitate the cerebrum by displaying a neuron and interfacing them through a weighted diagram.

3.3. Support Vector Regression

SVRs are unmistakable class of calculations, portrayed by utilization of pieces, nonappearance of neighborhood minima, meager condition of the arrangement and limit control acquired by following up on the edge, or on number of support vectors, and so forth.

4. Dataset

We have used the data bitstamp USD 1-min data from January 2012 to April 2019. The dataset is a 266775 row and 8 column dataset. Rows represent different data points. Columns represent parameters of the collected data. The columns are classified as:-

- Timestamp- Represents time at which the values are recorded.
- Open- denotes the opening price in the stock market.
- High- Represents the highest value incurred within opening and closing price.
- Low- denotes the lowest value incurred within opening and closing price.
- Close- denotes the closing price in the stock market.
- Volume of Bitcoin Traded- Represents the volume of bitcoin traded.
- Volume of Bitcoin Traded in price- Represents the actual price of the traded volume.
- Weighted price- Represents the average price corresponding to a timestamp.

Context

This data set contains minute wise data on bitcoin values

Content

Time Period: January 2012 – April 2019 (7 years 4 months)
Granularity: Minute Wise

5. Predictive analysis

For the predictive analysis of the bitcoin using the ML algorithms, RMSE value is used for evaluating the performance

RMSE can be defined as:

$$\text{RMSE} = \sqrt{(\text{p} - \text{a})^2}$$

Where p is the predicted value and a is the original value.

The values of mean squared root errors on test obtained on the three machine learning algorithms: Support Vector Regression, Linear Regression, Neural Network Regression (Table 1). Graphs are also obtained by plotting training data points with green and test data points with red. (Figure 1-3)

Table 1 RMSE

S. No.	Algorithm Used	RMSE value
1.	Support Vector Regression	1.279
2.	Linear Regression	1.176
3.	Neural Network Regression	3.71

5.1. Support Vector Regression

In figure 1, it can be seen that the green and red points are clearly distinct at certain data points. This indicates significant errors in the predictions made by the model.

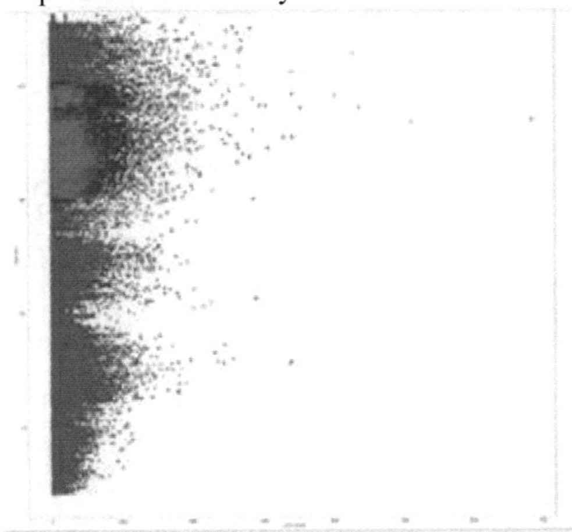

Figure 1. Mean squared root errors on Support Vector Regression

5.2. Linear Regression

In figure 2., it can be seen that the green and red points are not much distinct. This indicates quite low error in the predictions made by the model. However, there is significant error at some data points.

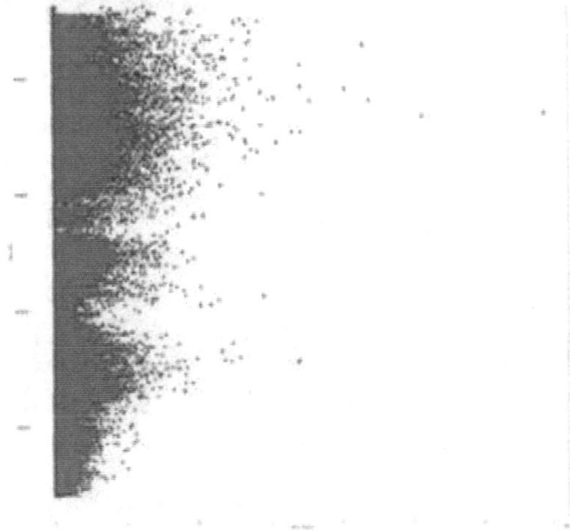

Figure 2. Mean squared root errors on Linear Regression

5.3. Neural Network Regression

From figure 3, the distinction between green and red points is between the observations of 1st and 2nd model, indicating low error compared to 1st model but higher than 2nd model.

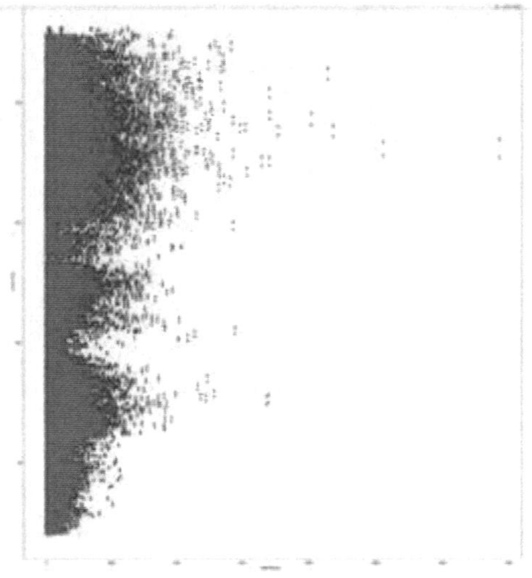

Figure 3. Mean squared root errors on Neural Network Regression

By the analysis of Table 1 and figure 1-3, it can be said that linear regression predicted the value of the bitcoin most accurately among the three algorithms.

6. Conclusion

In this paper, machine learning models – LR, NNR and SVR have been used to predict the bitcoin value. To compare the algorithms, root mean squared error value is used as the accuracy measure. Results show that LR performed best among the three algorithms.

In this paper, only three algorithms have been used.. Other various ML algorithms can be explored in predicting the value of the bitcoin. Also, here only one dataset is used for the training and the testing purpose as well. Multiple datasets can be used to get more accurate results.

References

[1] M. Günay and T. Ensarı, "Predictive churn analysis with machine learning methods," 2018 26th Signal Processing and Communications Applications Conference (SIU), Izmir, 2018, pp. 1-4

[2] A. Kanawaday and A. Sane, "Machine learning for predictive maintenance of industrial machines using IoT sensor data," 2017 8th IEEE International Conference on Software Engineering and Service Science (ICSESS), Beijing, 2017, pp. 87-90.

[3] A. Charleonnan, T. Fufaung, T. Niyomwong, W. Chokchueypattanakit, S. Suwannawach and N. Ninchawee, "Predictive analytics for chronic kidney disease using machine learning techniques," 2016 Management and Innovation Technology International Conference (MITicon), Bang-San, 2016, pp. MIT-80-MIT-83.

[4] P. S. Rajawat, D. K. Gupta, S. S. Rathore and A. Singh, "Predictive Analysis of Medical Data using a Hybrid Machine Learning Technique," 2018 First International Conference on

Secure Cyber Computing and Communication (ICSCCC), Jalandhar, India, 2018, pp. 228-233.
[5] S. K. Pushpa, T. N. Manjunath, T. V. Mrunal, A. Singh and C. Suhas, "Class result prediction using machine learning," 2017 International Conference On Smart Technologies For Smart Nation (SmartTechCon), Bangalore, 2017, pp. 1208-1212.
[6] M. Z. Hassan, M. E. K. Ali, A. B. M. S. Ali and J. Kumar, "Forecasting Day-Ahead Solar Radiation Using Machine Learning Approach," 2017 4th Asia-Pacific World Congress on Computer Science and Engineering (APWC on CSE), Nadi, 2017, pp. 252-258.
[7] S. Karasu, A. Altan, Z. Saraç and R. Hacioğlu, "Prediction of Bitcoin prices with machine learning methods using time series data," 2018 26th Signal Processing and Communications Applications Conference (SIU), Izmir, 2018, pp. 1-4.
[8] S. Velankar, S. Valecha and S. Maji, "Bitcoin price prediction using machine learning," 2018 20th International Conference on Advanced Communication Technology (ICACT), Chuncheon-si Gangwon-do, Korea (South), 2018, pp. 144-147.
[9] B. Sakiz and E. Kutlugün, "Bitcoin price forecast via blockchain technology and artificial intelligence algorithms," 2018 26th Signal Processing and Communications Applications Conference (SIU), Izmir, 2018, pp. 1-4.

IOP Conf. Series: Materials Science and Engineering 1049 (2021) 012004 doi:10.1088/1757-899X/1049/1/012004

Identification of Fruits Using Deep Learning Approach

Deepali M Bongulwar
Department of Electronics and Communication, Sri Satya Sai University of Technology and Medical Science, Sehore (MP), India,Email:ddeepali2006@yahoo.com

Abstract: The paper intends to build a Model for the identification and classification of fruits using the concept of deep learning. The objective is to build an automatic system for feature extraction using convolutional neural networks. The system can sort the fruits. It can be put-to-use in checking the condition of fruits, whether they are fresh or not. The self-service system in the retail market can use it to identify fruits. The proposed system uses high quality 'ImageNet' dataset. The dataset consists of five different categories of fruit images. Dataset is very challenging. The model uses Convolutional Neural Networks to identify fruits from images. The accuracy obtained is 92.23%. Deep learning outperforms machine learning algorithms.

1. Introduction

There are various applications of identification and classification of fruits. Mostly, it is used in the identification of fruits in supermarkets and the robotic harvest in the agricultural field. It can be adopted in the identification of plant disease and species in the agricultural domain. In the supermarket, the items are identified physically by the sales associate, or through the self-service systems. The online retailer, Amazon built up a chain of grocery 'Amazon Go Grocery', which empowered the customer a just walk out shopping. The customers' can enter the store with the free app 'Amazon Pay' on their smartphone and shop like other supermarkets. The customer should have an Amazon account. The supermarket has an enormous number of digital cameras and sensors. Utilizing the concept of deep learning and computer vision algorithms, the technology recognizes the items the customers' pick. You need not stand in a line or checkout. The charges of picked items are deducted from the account of the customer [1].This paper deals with the development of the identification system. An image classifier is trained to identify different images of fruits. The motivation behind the use of this technology is to enhance the identification process done by the self-service systems in the supermarket. Here, explicitly the system depends on the convolutional neural networks (CNNs).CNNs have a large number of neural network layers that can progressively learn different features [2].CNNs are accustomed either to collect or count the fruits. The models using deep learning have mostly employed a previous couple of years for autonomous feature extraction. These models are most powerful in feature extraction. [3].They have obtained exceptional accuracy in several tasks like the classification of images, recognition of objects, and speech recognition [4] [5].A dataset of five categories, constituting 4,760 images is employed to check the functioning of the model. The greatest accuracy of classification achieved on the dataset is 92.23%.

The first part of the paper explains the objective of the paper, its need and the related work. Secondly, it describes the achievements in deep learning followed by the methodology. It describes the structure of CNN Model used and the training dataset as well as the obtained performance. Finally, it ends with the conclusion and future scope.

2. Deep Learning

Deep learning specifies that a computer figures out how to perform classification from images without manually extracting the features [2]. The deep learning models have accomplished the state-of-art accuracy beyond human-level capabilities. The training of deep learning models is done by utilizing a

huge amount of labeled data. Convolutional neural network architecture constitute numerous layers. The word 'deep' is used to refer more number of hidden layers. The deep learning is similar to machine learning that utilizes numerous layers with nonlinear processing units [6] [7].
The features extracted from every layer are applied as input to the succeeding layer. CNNs are categorized as the deep learning algorithms. These networks contain many convolutional layers and some fully connected layers. They additionally utilize pooling layers. The different deep learning architectures (DLA) like Recurrent Neural Networks (RNN), Convolutional Neural Networks (CNN) and Deep Belief networks (DBF) show state-of-the-art results in numerous areas. Autoencoders can be used for generating handwriting/speech/ images. Deep Belief Networks use Restricted Boltzmann Machine (RMB). (RMBs)/autoencoders can be used to train more deep networks. CNNs gives exclusively best accomplishment in object recognition. This made us choose deep learning to identify fruits from images. Deep neural networks perform much better than any machine learning algorithms. Furthermore, deep neural networks, particularly convolutional neural networks have demonstrated to get extraordinary outcomes in the recognition of images [8].

3. Methodology

This section outlines the methodology and dataset used to develop the Fruit Identification System and presents the system that has been used.

3.1. Convolutional Neural Network

Here, a convolutional neural network has been used. It consists of convolutional layers, batch normalization layers, pooling layers, ReLU layers and fully connected layers. Convolutional layers comprise kernels of different size. The depth of the kernels is similar to the given image. The neurons of the kernel are connected only to the local regions of the input image called the receptive field. Every neuron has its own weights. The kernels slide horizontally and vertically on the input image and result in a two-dimensional feature map. The parameter, stride decides how a kernel can slide on the input images. Pooling layers turn down the spatial dimensions of the input images and lessen the number of calculations. Thus, pooling layers control the overfitting of the Model. Here, the pooling layers have filters of size 2 x 2 and a stride of 2.The size of the image is changed to one-fourth of the input. ReLU (Rectified Linear Unit) layer is the activation function, max (0, x).These layers make no changes in the size of the network, but increase its nonlinearity. Finally, neural networks have fully connected layers. Every neuron of a fully connected layer is connected to the resulting output of the past layer [8].
This is generally the outermost layer of the network. Softmax layer is utilized for anticipating a category from numerous categories. The input comprises images of size 224x224x3 pixels. The MATLAB software with Neural Network Toolbox is utilized to create this convolutional neural network.

3.2. The Architecture of the proposed CNN Model

The proposed framework uses a deep learning architecture. The model has 41 CNN layers. The figure 1 below shows the architecture of the CNN Model. It comprises feature extraction and classification. The input images are cropped to remove any unwanted information. All the images are resized to 224X224X3 (3 corresponds to three color channels Red, Green, and Blue).The feature extraction is done by applying convolution and pooling layers number of times. The model shows six consecutive blocks and fully connected layers. First two blocks contain two consecutive convolution layers and max pooling layers and later on four consecutive layers of convolution. Each block has rectified linear unit (ReLU).This is non-linear activation function [9]. Batch normalization layers are applied after convolutional layers and before ReLU layers. This decreases the training time and reduces the sensitivity to network initialization. Here, the size of filters is fixed i.e. 3×3 for convolution and 2×2 for max-pooling are utilized. The CNN layers shows the feature maps of the images at different levels. The last layer is the classification layer. Here, the two-dimensional feature maps are flattened into one-dimensional feature vector. This resulting feature vector is applied as input to the outermost layer. The outermost layer is a fully connected layer of neurons with the number of categories equal to the number of input categories. The output layer uses softmax function. It finds the probability for each category and outputs the class with maximum probability. The dropout layers are added before the fully connected layer. There are four dropout layers. The dropout layers are used for regularization of the network and thus avoids overfitting.

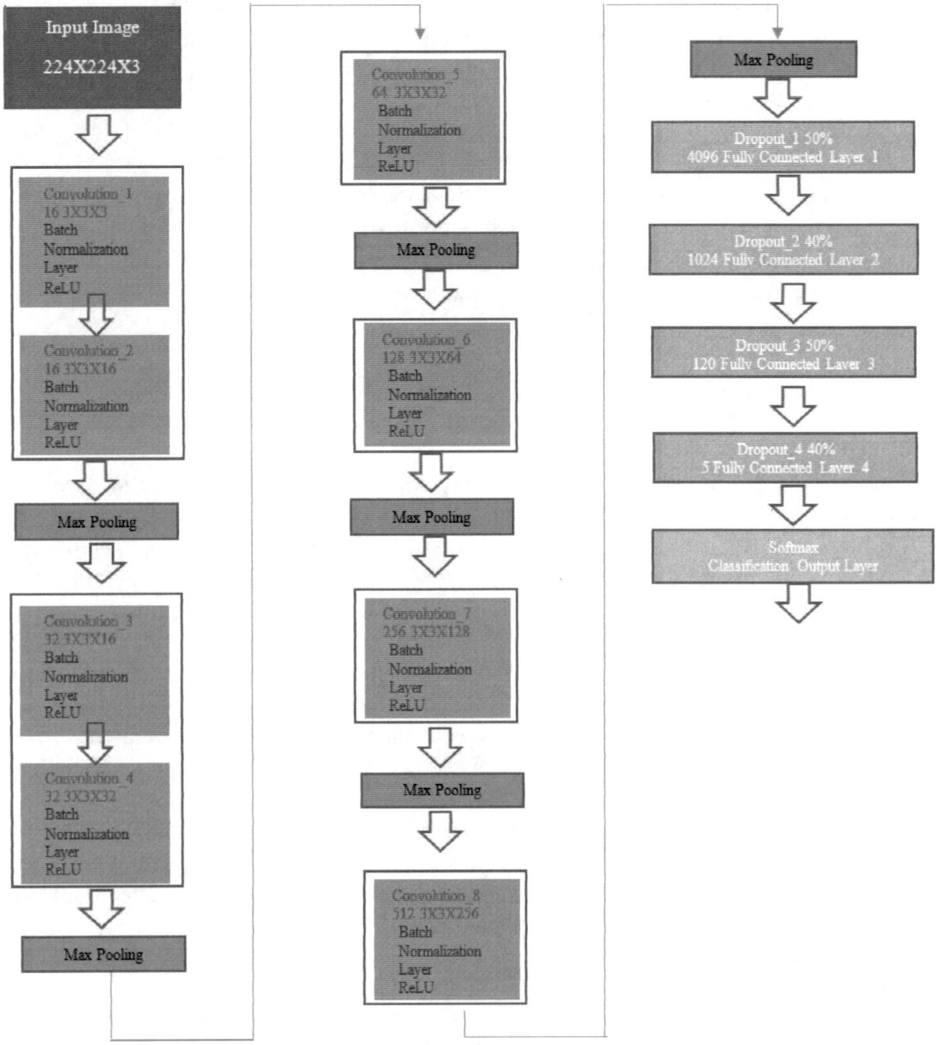

Figure 1. The Architecture of the proposed CNN Model

3.3. Dataset

All the images are downloaded from the 'ImageNet', free open dataset. Dataset consists of images of five different categories as shown below in table 1. The five categories are Apple, Banana, Grapes, Litchi and Mango. The RGB images, with three color channels R, G, and B are utilized in the dataset. The randomly selected sixteen images of the used dataset are as shown in figure 2.The total number of images including all the five classes in the dataset are 4,760 as shown in table 1.

The images in the dataset contain fruits of the same class with a different size. The background of the images is not homogeneous. The dataset has different poses of the same types of fruits. Fruits are with different positions and views like top view, side view, different background, half cut, sliced on the dish, cut in pieces, half-eaten, showing the seed, partially occluded. Sometimes fruits are fresh, sometimes rotten, or in bundles. Some images are with poor illumination, with different light effects, covered with snow or net, decorated, painted, along with leaf, on trees, inside a box or a plastic bag etc.

Black/white images are also considered. The dataset is very challenging. This is a very important factor in the proposed system.

Table 1. Input dataset

Label	Count
Apple	977
Banana	918
Grape	995
Litchi	964
Mango	906

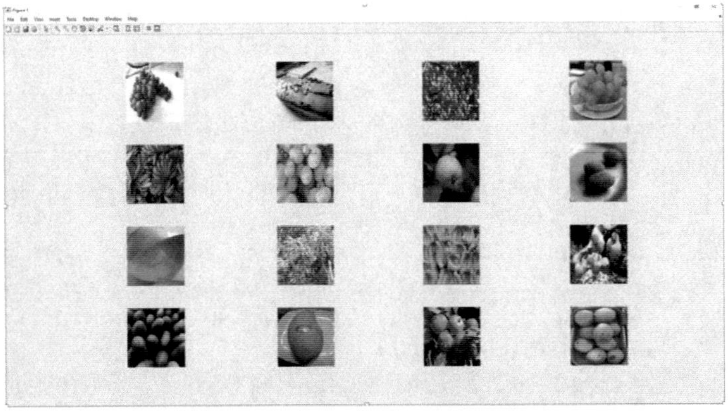

Figure 2. Sample Input Images

4. Experimental Results and Discussion

The images in the dataset are preprocessed and resized to 224x224x3 and cropped to remove any additional information. The dataset is divided into training and validation datasets in which 90% of the images are trained and 10% are validated. The optimizer, stochastic gradient descent (SGD) with piecewise learning is used. The minibatch size is 128. The initial learning rate is 0.01. The learning rate drop factor is 0.02. The learning rate drop period is 40 epochs. The images in the training dataset are augmented. The 128 samples in every minibatch are augmented as below:

-Randomly rotated between 0^0 and 90^0

-Horizontal reflection

-Horizontal and vertical translation

The proposed CNN model has achieved classification accuracy of 92.23%. The training and validation progress for each iteration of the epoch is shown graphically in figure 3. The labels on the output images show that they are accurately identified. The output images in figure 4 are pieces of fruits, still they are perfectly identified. The model is robust and gives very good accuracy.

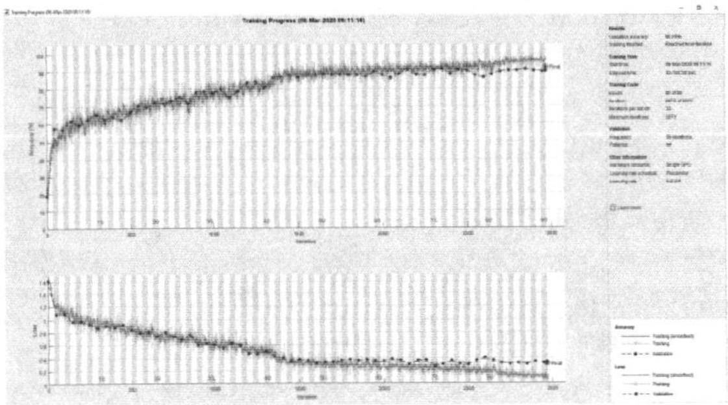

Figure 3. Training Progress versus Epoch Number

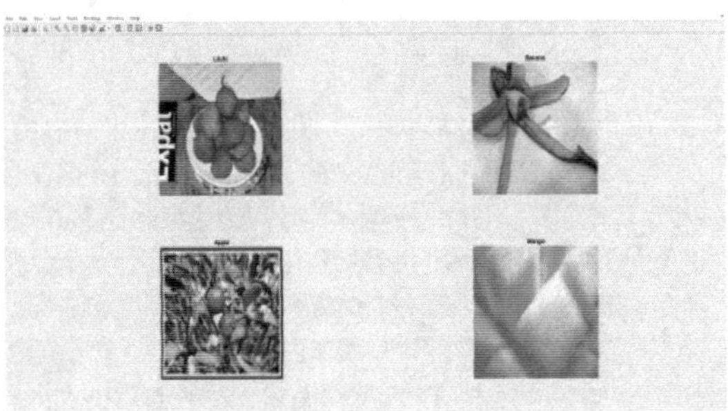

Figure 4. Resulting Output Labeled Images

5. Conclusion

The deep learning model for fruit identification and classification is developed in the paper. The paper has presented a system that develops an autonomous identification of fruits by the self-service system in the supermarket. The CNN Model has accomplished outstanding accuracy on the dataset.

IOP Conf. Series: Materials Science and Engineering 1049 (2021) 012004 doi:10.1088/1757-899X/1049/1/012004

As a future scope, the model can be used to train more variety of fruits. It can also examine the impact of various parameters like activation function, pooling function and optimizers.

References

[1] Frida Femling, Adam Olsson and Fernando Alonso-Fernandez 2018 Fruit and Vegetable Identification Using Machine Learning for Retail Applications *14th Int. Conf. on Signal-Image Technology & Internet-Based Systems (SITIS)* pp 9-15

[2] Y LeCun, L Bottou, Y Bengio and P Haffner 1998 Gradient-based learning applied to document recognition *Proc. of the IEEE* vol 86 pp 2278-2324

[3] X Yuan et al 2018 Deep Learning-Based Feature Representation and Its Application for Soft Sensor Modeling with Variable-Wise Weighted SAE *IEEE Transactions on Industrial Informatics* **14** pp 3235-43

[4] Z Gao, L Wang, L Zhou and J Zhang 2017 HEp-2 Cell Image Classification with Deep Convolutional Neural Networks *IEEE J. Biomed. Heal. Informatics* vol 21 pp 416-28

[5] T Kooi et al 2017 Large scale deep learning for computer-aided detection of mammographic lesions *Med. Image Anal* **35** pp 303–12

[6] Stuart J Russell, Peter Norvig 1995 *Artificial Intelligence-A Modern Approach* ed Egnlewood Cliffs (New Jersey: Simon and Schuster Company and Prentice-Hall) pp 79-80.

[7] Jurgen Schmidhuber 2015 Deep learning in neural networks: An overview of Neural Networks *National Centre for Biotechnology Information* **61** pp 85-117

[8] Horea Mureşan, Mihai Oltean 2018 Fruit recognition from images using deep learning *Acta Universitatis Sapientiae Informatica* **10** pp 26-42

[9] M Shamim Hossain, Muneer Al-Hammadi and Ghulam Muhammad 2018 Automatic Fruit Classification Using Deep Learning for Industrial Applications *IEEE Transactions on Industrial Informatics* **15** pp 1027-34.

[10] G E Dahl, T N Sainath and G E Hinton 2013 Improving deep neural networks for LVCSR using rectified linear units and dropout *IEEE Int. Conf. on Acoustics, Speech and Signal Processing (ICASSP)* (Vancouver BC Canada) pp 26-31

[11] Luis Santos Filipe N Santos, Paulo Moura Oliveira and Pranjali Shinde 2019 Deep Learning applications in agriculture: a short review *Robotics conf.* (Springer) pp 139-51

[12] Inkyu Sa, Zongyuan Ge, Feras Dayoub, Ben Upcroft, Tristan Perez and Chris McCool *DeepFruits:* A Fruit Detection System Using Deep Neural Networks

[13] Dan Claudiu Cire¸ Ueli Meier, Jonathan Masci, Luca Maria Gambardella and Jurgen Schmidhuber 2011 Flexible, high performance convolutional neural networks for image classification *Twenty-Second Int. Joint Conf. on Artificial Intelligence* pp 1237-42

[14] Rupesh K Srivastava, Klaus Greff and Jurgen Schmidhuber 2015 Training very deep networks, Advances in neural information processing systems *Twenty-Eight Int. Conf. on Neural Inf. Processing Systems* (Montreal Canada) pp 2377-85

[15] Woo Chaw Seng and Seyed Hadi Mirisaee 2009 A New Method for Fruits Recognition System *IEEE Int. Conf. on Electrical Engineering and Informatics* **1** pp 130-34

[16] Yudong Zhang, Shuihua Wang, Genlin Ji and Preetha Phillips 2014 Fruit classification using Computer vision and feedforward neural network *Journal of Food Engineering* pp 43-67

Water Feature Extraction, Enhancement and Change Detection of Multi-Temporal Satellite Images using MNDWI2-PCA

M Hemalatha
ECE department, Chadalawada Ramanamma Engineering College, Tirupati, A.P, India.
maddihemalatha@gmail.com

Abstract- Water feature extraction is a challenging task in remote sensing. In this research work, a new water index is implemented for easy identification of water pixels. The Area of interest is extracted with desired shape file. Here water bodies from kalahasti region are extracted, which is in Chittoor district. The Water indices are used to identify water pixels from Landsat-8 image, which has high spectral resolution. This image is multi-spectral image comprising of eleven bands. Interactive supervised classification is implemented for segmenting the satellite image. The image is classified into two categories i.e. water bodies and non-water bodies. Then MNDWI2-PC (Normalized Difference Water Index2 -Principal Component) is applied to LANDSAT-8 image. Then this image is segmented into water bodies and non-water bodies. Finally accuracy assessment is carried out by confusion or error matrix. Quantitative parameters such as Overall Accuracy (OA), Kappa Coefficient (KC), Overall Kappa Coefficient (OKC) User's accuracy (UA), Producer's Accuracy (PA), and F1score (F-Measure) are calculated for this multi-spectral satellite imagery. The algorithm reduced misclassification of water pixels with urban pixels, vegetation and other land covers. The algorithm outperforms in terms of quantitative performance metrics.

Keywords –Normalized Difference Water Index –Principal Component, Modified Normalized Difference Water Index -Principal Component, error matrix, and accuracy assessment.

1. Introduction

Many water indices have been used for water feature Extraction. McFeeters proposed Normalized Difference water index (NDWI) to extract ponds and lakes related to wetlands [1]. Xu proposed modified normalized difference water index (MNDWI) for identifying water pixels. Ji proposed a new method where he replaced band 4 with band 5 in MNDWI and suggested that, this method was more stable [2]. Automated Water Extraction Index with no shadow ($AWEI_{nsh}$) and Automated Water Extraction Index with shadow ($AWEI_{sh}$) are the two WI's (Water Indices) proposed by Feyisa [3]. Water pixels are mixed up with background noise and According to time and location of image acquitted, the threshold values are varied [4]. PCA has two merits one is dimensions of data set are reduced and second is noise is reduced. Without PCA (Principal Component Analysis), the author in [5] calculated WI's. Change detection with NDWI-PC has got good accuracies, proposed by Komeil Rokni [6]. Fusion of remote sensing indices gave good results for water pixel extraction [7]. The Proposed method gave better results in water feature extraction and qualitative performance metrics.

2. Materials And Methods

Kalahasti area from Chittoor district in Andhra pradesh is used for water bodies' detection and analysis. To analyze water bodies from satellite images, ArcGIS 10.3 software is used. The total image Resolution is 7631×7801. The satellite images are pre-processed by nearest neighbor interpolation algorithm. Area of Interest (AOI) is extracted from 143/50 (row/path). Here, AOI is kalahasti, which is extracted from Chittoor district. Geo-processing clip is applied for shape file. Then Raster clip is applied for satellite image and shape file, to extract desired area.

The PCA algorithm is used to achieve unco-related data set. In satellite imagery each band corresponds to different wavelengths. With PCA, the bands are reduced to give unique information. The merits of principal component analysis are to reduce the dimensionality and to reduce the noise. The PCA is also used for enhancement of satellite images. Finally, Interactive supervised classification is used for identifying land-covers. All the evaluation performance metrics are calculated with the help of confusion matrix.

For validating satellite data, 25 ground truth points (GTP) or tie points in each class are selected in satellite data. Total 50 tie points are selected two land covers. Shape file is created for these two classes with 50 tie points. Then, shape file is converted into raster. This resultant image is combined with classified data. Finally pivot table is generated, which gives confusion matrix for classified data and reference data. While creating shape file, the input satellite data and shape file data should be in same coordinate system. Otherwise pixels in the satellite data are misclassified. Quantitative performance metrics such as OA, KC, OKC, PA, UA, and F1Score are calculated. The algorithm for water extraction and enhancement is shown in Figure 1.

Water features are easily identified by Modified Normalized Difference Water Index 2 -Principal Component (MNDWI2-PC). The misclassified with this algorithm are minimized. The edges are detected well and the image is enhanced with the PC.

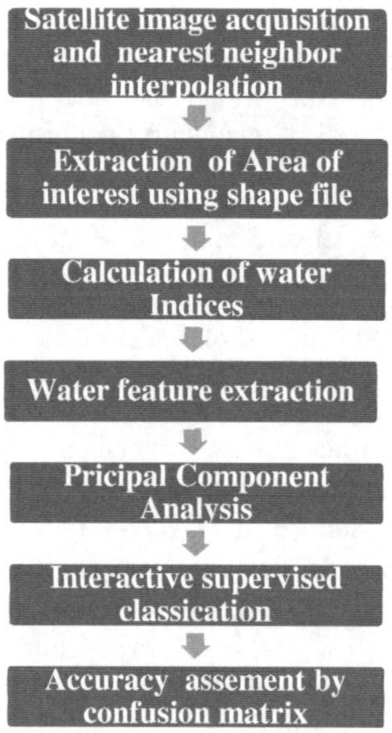

Figure 1. Flow chart for water feature extraction

2.1 Data set

The data set is downloaded from USGS (United States Geological Survey) earth explorer. These images are collected from satellite LANDSAT-8 OLI (Operational Land Imager). Table 1 describes Landsat-8 0LI characteristics. Table 2 presents various bands of Landsat-8 0LI. In this research letter, extraction of water is main scenario. So band 2, band 4, band 5, and band 7 are selected for water feature extraction. They gave better results compared to other spectral bands.

Table 1. Landsat-8 OLI main characteristics

Acquired images dates	19/03/2019
Path/Row	143/50

Datum	WGS 84
Projection	UTM
Spatial Resolution	30m
File Format of Acquired images	Geo-Tiff
Total number of bands	11
Type of Sensor	OLI
Radiometric Resolution	16 bits
Temporal Resolution	16 days
Swath	190 km

Table 2. Landsat-8 bands description

Bands	Wavelength (μm)	Resolution (m)
Band 1	0.43-0.45	30
Band 2	0.45-0.51	30
Band 3	0.53-0.59	30
Band 4	0.64-0.67	30
Band 5	0.85-0.88	30
Band 6	1.57-1.65	30
Band 7	2.11-2.29	30
Band 8	0.50-0.68	30
Band 9	1.36-1.38	30
Band 10	10.6-11.19	100
Band 11	11.5-12.51	100

Here band 1 is coastal aerosol , band 2 is Blue, band 3 is Green (G), band 4 is Red (R), band 5 is Infra Red, band 6 is SWIR1 (Short Wave Infra Red) , band 7 is SWIR2, band 8 is panchromatic , band 9 is Cirrus, band 10 is TIRS1(Thermal Infra red), and band 11 is TIRS2. In this research work, the bands R, NIR, G, and SWIR are used for water feature extraction. The formulas for NDWI [6], MNDWI [6], and MNDWI2 are shown in equations 1, 2, and 3 respectively.

$$NDWI_{L8}=G_{L8}-NIR_{L8}/G_{L8}+NIR_{L8} \quad (1)$$
$$MNDWI_{L8}=NIR2_{L8}-R_{L8}/ NIR2_{L8}+R_{L8} \quad (2)$$
$$MNDWI2_{L8}= G_{L8}-SWIR_{L8}/ G_{L8}+SWIR_{L8} \quad (3)$$

PCA is calculated with Eigen value decomposition of data covariance matrix or singular value decomposition of data matrix.

PCA is dire necessary to decrease satellite image dimensions without any distortion. It eliminates noise in satellite imagery. The PCA is also used for feature extraction and enhancement of multi-spectral satellite images.

2.2 Study Area

The proposed water feature extraction algorithm is implemented on kalahasti region in Chittoor. Chittoor district is in Andhra Pradesh (India). It has longitude of $79^0 10^{'} 03^{"}$ and latitude of $13^0 21' 72''$. It belongs to Rayalaseema region of AP.

3. Results And Discussions

The False color image (FCC) is shown in Figure 2. This image is the resultant of adding 4, 3, and 2 bands. Figure 3 explains NDWI-PC classification after applying principal component to NDWI (2014). Figure 4 explains MNDWI-PC classification after applying principal component to MNDWI

(2014). Figure 5 describes MNDWI2-PC classification after applying principal component to MNDWI2 (2014).

The False color image (FCC) for 2018 is shown in Figure 6 and NDWI-PC classification after applying principal component to NDWI (2018) is depicted in Figure7. Figure 8 describes MNDWI-PC classification after applying principal component to MNDWI (2018). Figure 9 explains MNDWI2-PC classification after applying principal component to MNDWI2 (2018).

Table 3 describes about surface area calculation for kalahasti image (2014). The area calculated for proposed method has got better value i.e. 52.21km^2. The same is validated with google earth reference pixels. Table 3 also describes about surface area calculation for kalahasti image (2018). The area calculated for proposed method has got better value i.e. 50.21km^2. The same is validated with google earth reference pixels.

Table 4 explains about accuracy assessment of kalahasti image in 2014. With this method, OKC is improved to 0.933 and KC is improved to 0.954. The method has got UA of 96.67 percent, PA of 96.67 percent and F1Meaure value is 0.949. OA is same with all the three methods. Table 5 explains about accuracy assessment of kalahasti image in 2018. With proposed method, OKC is improved to 0.933 and KC is improved to 0.932. The method has got UA of 96.67 percent, PA of 93.35 percent and F1Meaure value is 0.951. OA is 95 percent with proposed algorithm in 2018.

Table 5 describes graph for Change detection of water pixels for SriKalahasti in between 2014 and 2018. Change detection is performed by taking pixels difference in 2014 and 2018.

Figure 2. FCC for Sri kalahasti image (2014)

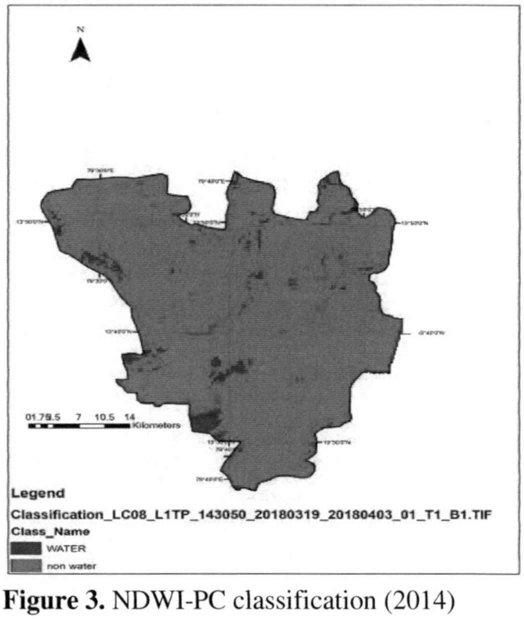

Figure 3. NDWI-PC classification (2014)

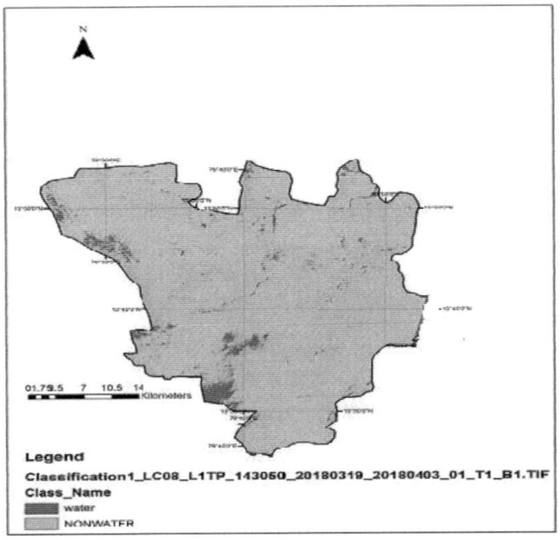

Figure 4. MNDWI-PC classification (2014)

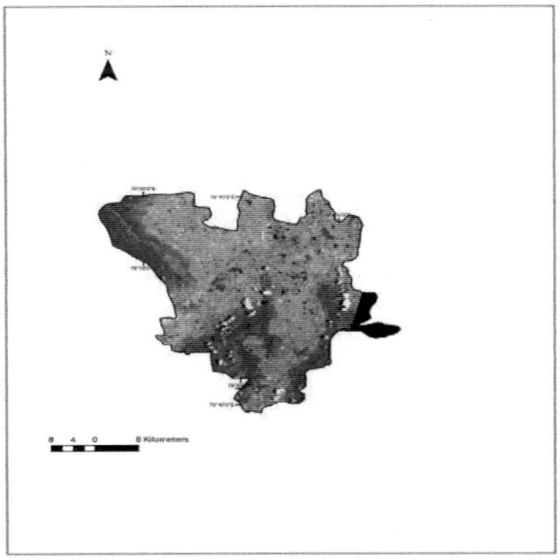

Figure 6. FCC for Sri kalahasti image (2018)

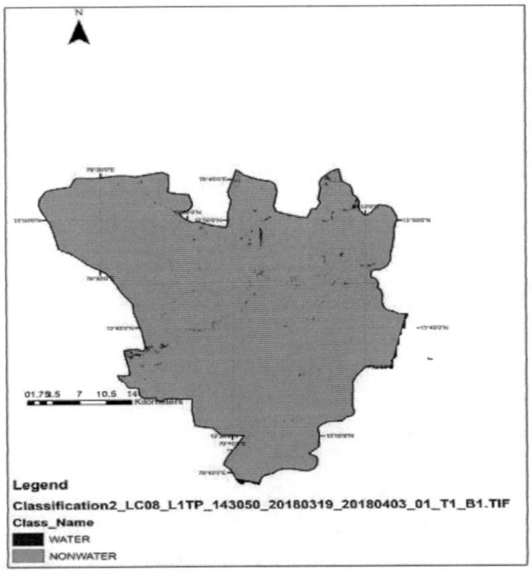

Figure 5. MNDWI2-PC classification (2014)

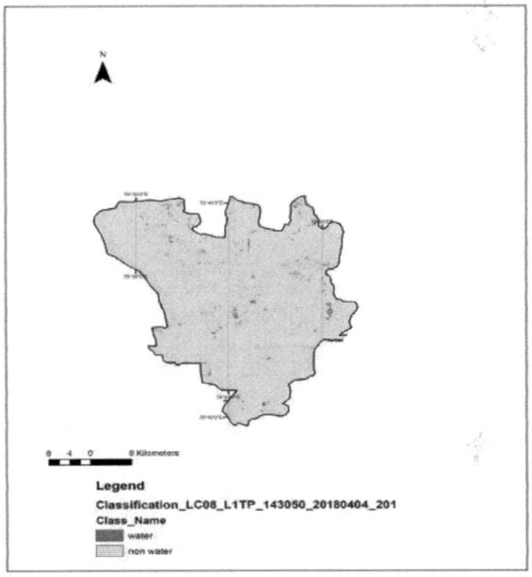

Figure 7. NDWI-PC classification (2018)

Figure 8. MNDWI-PC classification (2018)

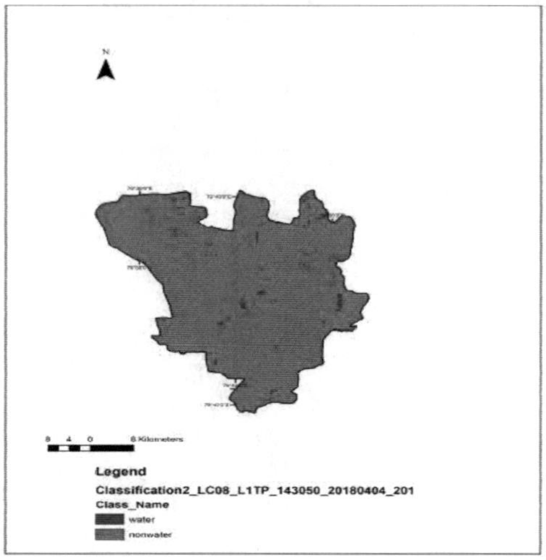

Figure 9. MNDWI2-PC classification (2018)

Table 3. Surface Area calculation of Srikalahasti in 2014 and 2018

Remote sensing indices	Pixel count	Spatial Resolution (m)	Area (Km^2)
NDWI-PC(2014)	59016	30	53.31
MNDWI -PC(2014)	62909	30	56.67
MNDWI2-PC (2014)	58006	30	52.21
NDWI-PC (2018)	58079	30	52.27
MNDWI-PC (2018)	62567	30	56.31
MNDWI2-PC (2018)	55797	30	50.21

Table 4. Accuracy assessment of Srikalahasti in 2014

Method	Area (Km^2)	Over All Kappa Coefficient (OKC)	Over All Accuracy (OA in %)	Kappa Coefficient (KC)	UA (%)	PA (%)	F-Measure
NDWI-PC	53.31	0.90	96.67	0.935	96.56	93.33	0.571
MNDWI-PC	56.67	0.931	96.67	0.954	93.75	96.77	0.956
MNDWI2-PC	52.21	0.933	96.67	0.956	96.67	96.67	0.949
Google earth reference data- 52.76 km^2							

Table 5. Accuracy assessment of Srikalahasti in 2018

Method	Area (Km^2)	Over All Kappa Coefficient (OKC)	Over All Accuracy (OA in %)	Kappa Coefficient (KC)	UA (%)	PA (%)	F-Measure
NDWI-PC	52.27	0.844	91.67	0.892	93.1	96.43	0.948

MNDWI-PC	56.31	0.9	95	0.934	96.55	93.33	0.949
MNDWI2-PC	50.21	0.933	95	0.932	96.67	93.35	0.951
Google earth reference data- 50.87 km^2							

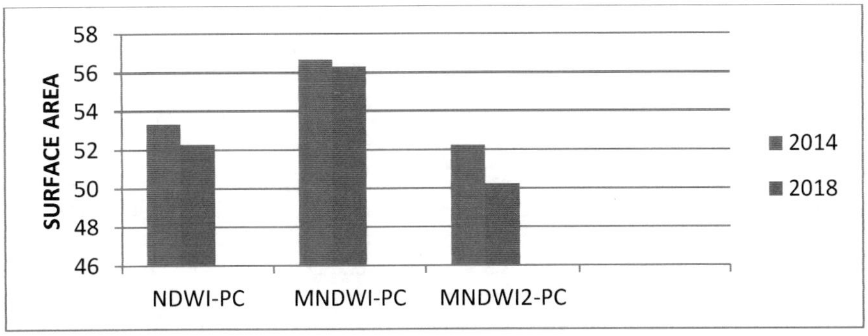

Figure 10. Graph for Change detection of water pixels for SriKalahasti in between 2014 and 2018

4. Conclusion

A novel algorithm is used to extract water from satellite imagery. With this method water bodies are identified properly. MNDWI2-PC is applied to the resultant image. Supervised classification is performed. Finally accuracy assessment has been performed with error matrix or confusion matrix. The proposed algorithm gave better results in terms of qualitative parameters such as OA, KC, OKC, UA, PA, and F-Measure. The method is able to extract water bodies in multi-spectral satellite imagery. Change detection has been performed with difference of four years.

Acknowledgement

The authors appreciate USGS for providing LANDSAT-8 ETM+ images. The authors would also like to express their appreciation to S. V. University, C.O.E (centre of excellence) laboratory, Tirupati, for providing all facilities and assistantship and necessary software's for our research.

References

[1] S. McFeeters, "*The use of the Normalized Difference Water Index (NDWI) in the delineation of open water features",* Int. J. Remote Sens., **Vol. 5**, pp. 1425–1432, 1996.

[2] H. Xu, "*Modification of normalised difference water index (NDWI) to enhance open water features in remotely sensed imagery*" Int. J. Remote Sens. **Vol. 27**, No. 14 pp. 3025–3033, 2006.

[3] G.L. Feyisa, H. Meilby, R. Fensholt, and S.R. Proud, "*Automated water extraction index: A new technique for surface water mapping using Landsat imagery*", Remote Sens. Environ., **Vol.140**, 23–35, 2014.

[4] L. Ji, L. Zhang, and B. Wylie, "*Analysis of dynamic thresholds for the normalized difference water index*", Photogramm. Eng. Remote Sens., **Vol. 75**, No. 11, pp.1307–1317, 2009.

[5] L. Yongxue, S. Chao, "*Automated Extraction and Mapping for Desert Wadis from Landsat Imagery in Arid West Asia*", Remote Sens., **Vol.8**, No.246, pp.1-23, 2016.

[6] Komeil Rokni, Anuar Ahmad, Ali Selamat, and Sharifeh Hazini, "*Water feature extraction and change detection using multi temporal lansat imagery*", Remote Sens., **Vol.6**, 4173-4189, 2014.

[7] M.Hema Latha, S.Varadarajan, "Feature enhancement of multispectral images Using vegetation, water, and soil Indices Image Fusion, ISMAC-CVB, Springer Conference, May 2018 .

IOP Conf. Series: Materials Science and Engineering 1049 (2021) 012006 doi:10.1088/1757-899X/1049/1/012006

Prediction of Learner's Performance in Adaptive E-Learning System using Learning Analytics

Soni Sweta[1], Shalini Mahato[2], Laxmi Kumari Pathak[3]

[1]*Department of Computer Science & Engineering AmityUniversity Jharkhand, Ranchi, soni.sweta16@gmail.com*

[2] *Department of Computer Science & Engineering Amity University Jharkhand, Ranchi, smahato@rnc.amity.edu*

[3]*Department of Computer Science & Engineering Amity University Jharkhand, Ranchi, lkpathak@rnc.amity.edu*

E-mail: soni.sweta16@gmail.com

Abstract. The main objective of Learning Analytics is to collect, interpret and investigate the information for setting proper co-relation to improve the students' learning experiences. As the popularity and demand of learning analytics are on rise, higher education is continuously moving from offline to online E-Learning Educational System. It has been further promoted on account of COVID-19. Learning pedagogies, evaluation measures and feedback measures have also been changed accordingly. The aim of this paper is to discuss the efficacy of learning analytics in online educational system using e-learning platform. R programming language with GGPLOT2 is used as Visualization tool to focus on gain insights of adaptation of learning analytics. The results show the better computational performance in terms of predicting students to improve their learning achievements and mitigating risk of failures.

1. Introduction

Learning Analytics have huge recognition and acceptance among teachers and analysts in higher education[1]. Using Educational Data Mining (EDM) and Learning Analytics (LA) in E-learning System, we can develop learner's centric model that is useful in taking decision and making strategy for continuously ensuring improvement of the organization[2].The primary idea of learning analytics is very interesting. Its ideas and methods can used by educators to predict their learner's outcomes by using EDM in advance and take right decision. It is very useful for schools, colleges and universities for taking better decisions, accumulating huge data and its information and applying analytical tools to maximize decision making capability in terms of resource utilization, students' achievements and outcomes[3]. Learning Analytics is an interdisciplinary research area because of its linkages with virtual learning and digital teaching that is useful in EDM, e-learning system and its statistics. The huge data generated during online teaching sessions are gathered, examined and then analysed the patterns and behavioural trends to prepare feedbacks on behavioural traits[4].

These mined "noteworthy knowledge" supports the teaching and learning and recommends personalized and customized information. Online virtual education systems support the future growth and development in virtual university and adapt to handle the uncertainty like COVID-19 pandemic[6].Higher education needs to enhance infrastructure and networking; so that more accountable financing models and guaranteed educational system can be built in view of operational effectiveness, accepting adoptive changes and competitive at all levels. E-learning education not only manage extravagant overheads, but also understand the actual needs of students who comprehend the requirements of the organization[5].

In this challenging digital era, more institutions use techniques for Learning Analytics so as to acquire discoveries on the research and development for analysing some interesting patterns and finding for betterment an advancement of learning process. Further, it is a suitable tools for analysing pattern in students learning behaviour to provide the support and feedback to educators or tutors[7].

The utilization of Learning Analytics in Universities is to improve the inter institutional relationship, co-operation and collaboration, which motivate for development of agenda for large group of students and teacher community[8].

This paper's Second section is about review literature on LA in higher education, explained the proposed methodology in third section and discussion followed by conclusion discussed in fourth and fifth section, respectively.

2. Literature Review on Learning Analytics in Higher Education

We thoroughly examined various papers on LA in the related research work and gained insights into the pattern and trend. The description about LA and its increasing popularity has discussed in Horizon Report 2012 by Johnson et al[9]. as "the estimation, gather, examination and visualizing of data and information about students and their usefulness for ability of comprehension and upgrading learning and the environments in which it happens". In another way, we can express as "`the utilization of intelligence, learner-centric information, and analysis models to find new patterns among hidden information to predict the learning outcomes and give the feedback" in a journal[10]. A model was discussed for the identification of students at risk based on their risk parameters, and mitigation of students risk to enhance learning awareness and benefit both students as well as teachers[11]. This paper discussed the current scenario of LA and revealed its effectiveness in higher education [12]. In this paper, dynamic testing for the development of cognitive skills and its prediction is very important in comparison to the existing set of learner's characteristics[7]. In this paper, learners' behaviours were identified and matched with their profile and their attributes by developing a model on the basis of algorithms similarity computation and Jaccard coefficient which helped in the improvement and optimization e-learning[13]. A review of the developments in the field of educational data mining, learning analytics are explained in big data scenarios briefly and with some models used are: multilayer perception (MLP)network, multiple linear regression, radial basis function (RBF) network and support[8]. The authors have developed a learning analytics model in which dynamic interaction among different stakeholders is proposed with the help of visual analytics. It mainly focused on personalized learning and supported to increase retention rate [6].

3. Methodology and Execution

3.1 Learning Process

Authors Campbell and Oblinger in 2007[14] said that , there are five steps of learning analysis process , shown in Fig. 1. Capturing real time data and gathering information from various sources like virtual learning environment, LMS. It contains Blog, Forum and Chat etc like features. There is a reporting mechanism in which the collected data are used to make a model to enhance the students' progress. Before making any prediction, first we need to visualize the data set which gives the relationship among variables then we make a model based on high impact variables. Predicting, the information is utilized to recognize the predictor for learner's success and outcomes. Further, it is used for making decision of courses which utilized by the decision maker of the organization[15].Outcomes from data analysis process are useful for interventions in acting steps e.g providing extra support to those learners who may have more chances of failure or at risk in the course.

In refining, collected information are used for continuous support and improvement in teaching learning process. All the stakeholders need to discuss the idea and opportunities on how to incorporate all these possibilities in their research and development. Even though the main aim of LA is to focus on learning process, but its result shows the benefits of all types of stakeholders associated with the system.

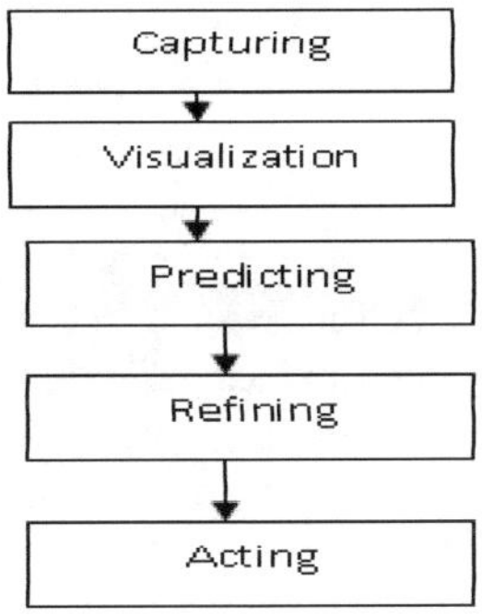

Figure 1. The five steps of the Learning Analysis process.

3.2 Different Stakeholders
According to Romero and Ventura in 2013[9] there are four classes of stakeholders which were divided on the basis of their own perspective, aim and objectives. Learners Group supports the learner with adaptive feedback, personalization, customization of the responses according to their preference, the learning performance improvement. Teacher Group, which understands students' learning process, reflects on teaching pedagogy and performance, understands cognitive and behavioral aspects. Researchers use the right educational data mining technique to find the hidden information which is beneficial for decision making strategy and evaluation of learning effectiveness in different challenging environment. Administrators, measure educational growth, do proper resource utilization, and generate educational offers.
In our research, we have considered a new type of stakeholder named as Mentor. Mentor bridges the gap between learners and teachers, parents and teachers and learners and institutions and counsels them to mitigate the learning gap. Mentor does mentor all the time during the course of their mentee. It will improve the overall personality of mentee and their performances.

3.3 Task to be performed - Algorithm and Implementation
We consider some factors in e-learning system which highly impact on achievement and outcomes of learners like using R programming language with DYPLYR package along with corr function. Then we get the significant value which show which variable have high impact over performance of students. Once we find then we compare it with the threshold value. Those students have less score then threshold value they will be at high risk. The threshold values will be decided by the faculty. It may vary course to course because of different courses have different difficulty levels as provided by the institution.
We have taken 140 students as a sample for e-learning courses. For academic achievements we have analysed their marks obtain in quiz, project-based assignments, and participation in forum. For checking the engagement level, we analysed their behavioural patterns like time stamp, no. of attempts, time spend on completing the session. The academic marks obtained in multiple modes of exams and their behavioural values as mentioned in table 1 and LMS tool given below in figure2.

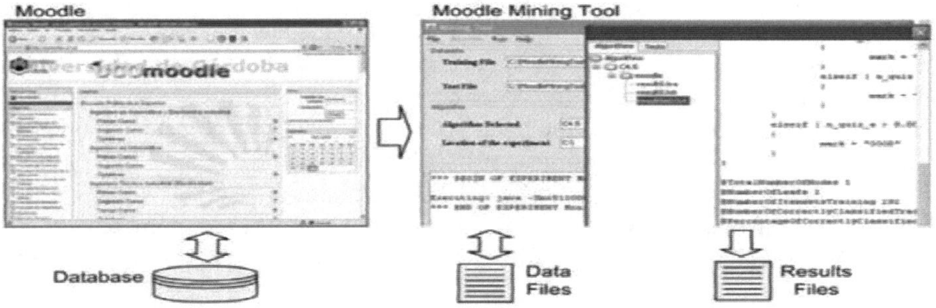

Figure 2. Moodle Virtual Learning Management Tool.
The SVM is a supervises learning method used to separate two group classification problems by classification algorithms. The support vector machines technique is developed by Cortes & Vapnik [15].It is used to separate two groups of data using hyper plane as defined by the support vectors. The separating hyper plane is decision boundaries which classify one type of data in one side and another type of data in another side. Many hyper planes can be drowning for the separation of data, but optimal line defines data in a better way. It helps in finding maximum and minimum of the margin between two classes of training data set to find out position of data in X-Y coordinates which is shown in figure 3.
Real world problems are generally non-linearly separated in which SVM separates data and minimizes misclassification. In this research work, there are huge amount of nonlinear data generated by LMS; so

SVM is the best way to classify these training datasets into two groups one is called risky and another one is non risky classes.

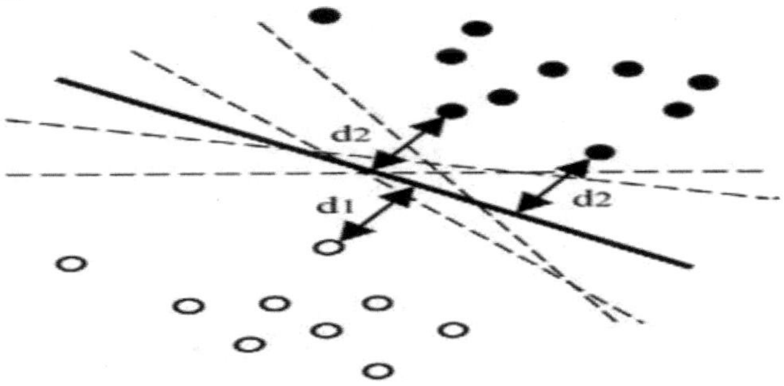

Figure 3. Support Vector Machine Technique.

4. Result and Discussion

In this research work, we have focussed mainly on time variant data features collected during e-learning course where 140 students were participated in C++ course. In which data related to learning objects like multiple choice test were conducted to measure theoretical knowledge, project work-Assignments were conducted to measure practical knowledge and semester end exams were conducted to see the differences in performance over a period of times. Participation in forum and chat also gave the interactive and engagement level/ sensitivity of students. Weightage of test is shown in table -1 20% project-work assignment ,10% forum/ group activity for engagement level ,20% for MCQ test, 30 % for end semester exams, 20% for demographic data in which we have 10% weight and 5 % weights for both communication medium and parental income. If students score more than threshold limit of 50% his or her chances will be higher to fall into non risky otherwise it will be risky. We can configure same data set with decision tree algorithm for comparing studies and find that SVM predicts with high percentage of accuracy as compared with decision tree.

$$\textbf{Accuracy} = \textbf{true\_predicted\_at\_ non risk} + \textbf{true\_predicted \_at\_ risk})/(\textbf{total\_number\_of\_learners}) \quad (1)$$

Table 1. Learner's features from data set for training and testing purpose

Based on	Features	Value range	Threshold value (%)
Time Dependent	Multiple choice question	0-30	15
	Project work assignment	0-20	10
	End semester exam	0-20	10
	Participation in forum /chat	0-10	5
Demographic	Area (Rural, Urban)	0-10	5
	Communication Medium (Basic, high, proficient)	0-5	2.5
	Parental Income (Low, Medium, High)	0-5	>0

Table 2. Learner's risk and non- risk cases examined

	Marks obtained in percentage			
Class	MCQ Test	End Term Test	Project-work assignment	Participation in Forum/ Chat
Risky	12	7	6	2
Non- Risky	21	16	18	6

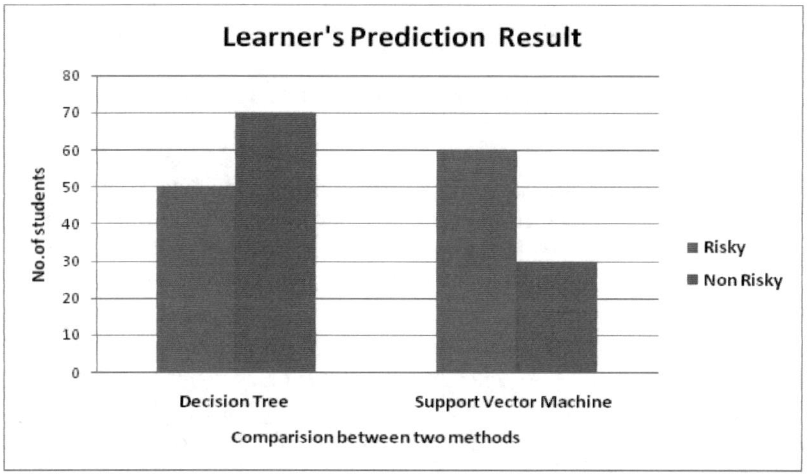

Figure 4. Accuracy results on the 140 students of the online course with both the method.

5. Conclusion

This paper presents a method of prediction of learner's performance in adaptive e-learning system using one of the machines learning techniques i.e.support vector machine. The proposed method used learner's demographic data along with data extracted from different actives performed by learners and learning process during the continuation of the course. It takes log data from the learning management system and finds patterns from the adaptive course to estimate where learners are lagging and how the proposed method helps them to enhance their performance dynamically.The SVM technique used to determine the prior demographic information and dynamics inputs during adaptive e-learning in terms of feedbacks, special assistances as personalised recommendations helped learners to prevent failures and enhance performance.

6. Future Aspects

In future other machine learning techniques may be applied to fine tune the model in terms of accuracy, precision, and sensitivity.

7. References

1. Jo IH, Yu T, Lee H, Kim Y." Relations between student online learning behavior and academic achievement in higher education: A learning analytics approach", Lect Notes Educ Technol. Emerging Issues in Smart Learning, Springer-Verlag Berlin Heidelberg, 2015,pp 275-287 DOI 10.1007/978-3-662-44188-6_

2. S. Sweta and K.Lal, "Adaptive and Personalized Intelligent Learning Interface (APIE-LMS) in e-learning System," International journal on Applied Engineering and Research: ISSN 0973-4562 Volume 10, Number 21 (2015) pp: 42488-92.

3. Peña-Ayala A. "Learning analytics: A glance of evolution, status, and trends according to a

proposed taxonomy", Vol. 8, Wiley Interdisciplinary Reviews: Data Mining and Knowledge Discovery. Wiley-Blackwell; 2018.

4. Lang C, Siemens G, Wise A, Gašević D,"Research A. Handbook of Learning Analytics SOCIETY for LEARNING", ISBN: 978-0-9952408-0-3,2017, DOI: 10.18608/hla17, 2017; www.solarresearch.com

5. Baker, R., & Siemens, G. (2014)."Educational Data Mining and Learning Analytics", In R. Sawyer (Ed.), The Cambridge Handbook of the Learning Sciences (Cambridge Handbooks in Psychology, pp. 253-272).Cambridge: doi:10.1017/CBO9781139519526.016

6. S. Sweta and K.Lal. "Personalised Adaptive Learner Model in E-Learning System Using FCM and Fuzzy Inference System" International Journal of Fuzzy Systems (IJFS), Springer, ISSN No 2199-3211, 2016 doi: 10.1007/s40815-017-0309-y.

7. S. Sweta and K. Lal, "Learner Model for Automatic detection of Learning Style Using FCM in Adaptive e-Learning System" International Organization of Scientific Research (IOSR) JCE eISSN -2278-0661, vol. 18, Issue 2, Ver.1 pp: 256–270, 2016.

8. Tsai Y-S, Gasevic D."Learning Analytics in Higher Education-Challenges and Policies", Proceedings of the Seventh International Learning Analytics & Knowledge ConferenceMarch 2017 Pages 233–242, doihttps://doi.org/10.1145/3027385.3027400.

9. Romero-Zaldivar V, Pardo A,"Monitoring student progress using virtual appliances: A case study",Computers & Education,Volume 58, Issue 4,2012,Pages 1058-1067,ISSN 0360-1315, https://doi.org/10.1016/j.compedu.2011.12.003.

10. Leitner P., Khalil M., Ebner M."Learning Analytics in Higher Education—A Literature Review", In: Peña-Ayala A. (eds) Learning Analytics: Fundaments, Applications, and Trends. Studies in Systems, Decision and Control, vol 94. Springer,Cham.,2017 https://doi.org/10.1007/978-3-319-52977-6_1

11. Larrabee Sønderlund A, Hughes E, Smith J." Student engagement and wellbeing over time at a higher education institution", A systematic review. Br J Educ Technol. 2019;vol.50(5),pp 2594–618. DOI:https://doi.org/10.1371/journal.pone.0225770

12. Sonderlund AL, Hughes E, Smith JR."The efficacy of learning analytics interventions in higher education-A systematic review", British Journal of Educational Technology, Vol 50 No 5 pp 2594–2618 2019, DOI: 10.1111/bjet.12720

13. Liang, Kun & Zhang, Yiying & He, Yeshen & Zhou, Yilin & Tan, Wei & Li, Xiaoxia.." Online Behavior Analysis-Based Student Profile for Intelligent E-Learning",. Journal of Electrical and Computer Engineering. 2017. 1-7. DOI: 10.1155/2017/9720396

14. Campbell J, DeBlois P,"Academic analytics: A new tool for a new era". EDUCAUSE Review, vol. 42, no. 4,pp 40-57, 2007; https://er.educause.edu/articles/2007/7/academic-analytics-a-new-tool-for-a-new-era

15. Cortes C, Vapnik V. Support-vector networks. Mach Learn 20(3), pp: 273–97, 1995 Sep;20(3), DOI: https://doi.org/10.1007/BF00994018

Bernoulli Vacation Scheduling (BVS) Model for Energy efficient intrusion detection in WSN

Gauri Kalnoor [1] and GowriShankar S [2]

[1,2]Department of Computer Science & Engineering, BMS College of Engineering, Bangalore, India.

E-mail: kalnoor.gauri@ieee.org (GauriKalnoor)

Abstract. Cutting-edge wireless sensor networks, security is the most important challenge that has to be considered. Most of the applications of WSN are vulnerable to attacks, as they are deployed in any kind of environment. So, the intrusion detection system is the second line of defense for WSN. The intruder should be detected in WSN more efficiently. We introduce the Bernoulli scheduling rule such that several trials are performed to get the detection of an intruder to be successful once. Since the sensor nodes are always active for detecting an intruder, the energy consumption becomes high. To reduce the consumption of energy, sleep/wakeup, also called small interval vacations are used for the sensors which are almost considered to be like the mechanism of fluid queues. In our proposed work, the vacations are introduced so that the energy used is minimized whenever the sensor is not required to be active. Thus, in our proposed work, since WSN is vulnerable to many harmful attacks, the Bernoulli scheduling rule is applied and the sensor nodes have a vacation state whenever the sensor need not be active, to save the energy consumed. The probability of detecting an intruder is calculated and performance is compared. We have applied a fluid queuing model for the vacations of small intervals in the sensor nodes to minimize the energy consumption and the obtained result is compared with the energy consumed when the vacations are not applied. Thus, the results are simulated for both energy consumed and the number of success rates of detection of an intruder, most importantly the comparison is analyzed with the past work of wsn.

Keywords: Vacation Policy, Bernoulli Scheduling, Energy consumption, WSN, queuing model, Intrusion Detection

1. Introduction

The sensor nodes are scattered in Wireless Sensor Network (WSN), which is an 'application-specific' network. It consists of tiny sensor nodes with energy limited and smart nodes of sensors. Every node in the network is capable of detecting a certain situation, processing some limited data, and communicating with each other [1]. WSN is mostly used in physical environments, which is collected by the small device called nodes; these nodes send information periodically to sink and retrieve information from the region in information-gathering apps and send it to the sink [2]. WSNs are often defined by resource-limited dense implementation in environments. However, the constraints are due to processing capacity, storage, and energy in particular because they are generally driven by batteries [3]. Recharging the batteries in the network of wireless sensors is sometimes impossible due primarily to the place of the nodes.

Many sensor nodes are merged to create a wireless network of sensors that differ tightly within the critical areas [4]. This sensor position cannot be reconfigured and this particular sensor's battery cannot be charged in critical are such as battlefields, thick forests, disaster-prone fields, and much more [5, 6]. These kinds of networks are used in many apps today, which vary in their goals and particular limitations. A sensor normally persists in these situations: an awake state and a sleeping state. A sensor in the waking state makes the size, some computations are performed and then the information is communicated to the center. The centers behave as a controller as well as a decision-maker. As sensors are energy-constrained (limited),a stringent requirement of the algorithm must be imposed encountering energy efficiency as the major factor. These imposed algorithms can be employed in the controller, i.e. fusion center.

The model of Intrusion detection [7] in WSN introduces some of the essential parameters like the range of sensing, node density, energy efficiency, accuracy, and so on. In the field of interest, the sensors are placed, where WSN can adapt to change in the topology of the network and the environmental conditions. The model can be designed for Intrusion detection based on a single node or multiple nodes of sensors and its sensing detection range, in WSN. One of the issues in WSN is the sensing coverage area and how accurately the monitored area is tracked or detected. Since, the analytical method is used to quantify the sensing range and other QoS parameters like buffer capacity, a period of sleep/wake of sensors, and also energy consumed during the process, other queuing models are used to limit the capacity of the buffer and schedule the vacation time of the sensor nodes.

Bernoulli's Rule

Bernoulli's law stems basically from the theory of energy conservation. This states that at all stages in a streamline, the sum of all forms of energy in steady flow is the same at all. It means that there is a constant sum of kinetic energy, potential energy, and internal energy. The proposed scheduling algorithm determines the nodes of the sensors based on their active, idle, and sleep modes. In order to assign time slots to all cluster participant nodes, the time ON and time OFF variables are computed for nodes that are either in sleep or active mode. The waiting time for sensor nodes in the cluster member is computed on their listening, sleeping, and activation level status. Additionally, it is presumed that sensor nodes regarded are of mobile nature and that the clusters are reorganized as a function of time.

2. RELATED WORK

In [8], the authors modified LEACH (MODLEACH) one of the most efficient wireless sensor networking protocols. In [9], the authors proposed a technique in which the hop's highest energy node becomes a cluster-head. The size of each cluster is, therefore, less than and/or equivalent to hops. The head of the cluster sends information to the other cluster head or sink with a tree-based minimum energy transmission algorithm owing to the transmission range restriction of nodes. The authors conducted simulations to compare its efficiency with standard routing protocol performance, such as direct and minimal transmission energy, and different energy levels.

In [10], the authors considered network cardinality estimation schemes that use the calculation of bit-wise maxima over strings as their basic aggregation mechanism. Therefore, under frequent assumptions, they evaluate how to effectively and efficiently select the parameters of the data generation process on the estimate, obtain the resulting Maximum Likelihood (ML) estimator and characterize its statistical performance as a function of communication and memory requirements. The authors have then compared the bitwise-max based estimator numerically with the lexico graphic-max based estimators and obtain insights into their comparative performance according to the real cardinality. The authors proposed a new, effective algorithm to calculate

The area covered by sensors in a region with transparent and opaque barriers and explores the issue of identifying and eliminating redundant sensors to enhance energy efficiency while maintaining network coverage. The suggested algorithm [11] utilizes methods of computational geometry and applies to both homogeneous and heterogeneous WSNs.

The authors [12] considered the redundancy of sensors in WSN and they performed an experimental study to better emphasize the importance of its use. They also introduced OER' Redundancy-based Energy Optimization,' a protocol that takes advantage of redundancy to save energy. Besides, they also stretch OER through a mechanism of fault tolerance. They demonstrate how OER coupled with the FTMO route performs traditional routing protocols that do not exploit redundancy through comprehensive simulations. The first break down in the power usage of the typical sensor node parts is discussed as the primary power preservation guidelines in WSN's. Then they presented a systematic and thorough taxonomy of energy conservation systems. Special attention has been given to promising alternatives, such as methods for energy-efficient data acquisition.

In [13], the authors, have proposed for WSNs an A-star algorithm, a fresh energy-efficient routing protocol (EERP). By forwarding data packets via the ideal shortest path, the suggested routing system increases the network lifetime. In terms of the next-hop sensor node's highest remaining energy, elevated connection quality, buffer occupancy, and minimum hop counts, the ideal route can be found. Simulation findings showed that the system suggested by the authors increase the lifetime of the network compared to the protocol A-star and Fuzzy Logic (A&F). In [14], the creators, proposed a vitality productive rest arranging component with a closeness measure for remote sensor systems (ESSM) to design sensors in either dynamic or rest mode so as to proficiently diminish vitality utilization. Initially, to adjust vitality utilization, the ideal span of rivalry is assessed to arrange all sensor hubs into a few groups. Second, a fluffy grid can be accomplished to assess the level of similarity as per the data accumulated by part hubs, and the relationship work dependent on the fluffy hypothesis can be set up to part the sensor hubs into different classifications.

3. MOTIVATIONS AND PROBLEM STATEMENT

The issues of the past work have been analyzed and such issues have been motivated for various reasons that include the random deployment in any region, with randomly distributed sensor nodes. The efficient deployment can be obtained in the compulsory sensing attention. However, specifically, given the region for monitoring in WSN, what makes us assure that each contact of the specified region is provided security by the number of sensors needed and covered at that point in the region? Thus, to be specific, the areas where the required sensors need to be deployed have to be identified such that the probability of detecting an intruder can be increased, and also the intruder will not be able to exceed the threshold set for the distance. Additionally, the issue of conserving energy is the challenging factor in WSNs, mainly due to the regulated energy supply by the sensor nodes, computational Capacity, and the availability of storage space. Therefore, it implies that the sensor network should have the capability of self-organization. Importantly, the required number of sensor nodes must be selected and placed effectively in the region of network coverage. The sensor's power is put in a state of ON/OFF so that the sensors will be in sleep or wake state to enhance the lifetime of WSN.

The proposed work considers the fluid queuing model for efficient utilization of the server in the network and also vacation scheduling using Bernoulli rule to improve the energy consumption of the sensor nodes. This model is termed as Bernoulli Vacation Scheduling (BVS) Model.

4. INTRUSION DETECTION MODEL

Intrusion detection designed in WSN is defined as the technique to detect the abnormal activity in the network in a certain range of sensing and also to detect the unauthorized anomalous attackers in motion or intrusions. The deployment of deterministic nodes is determined by analyzing sufficiently, before the deployment of the network. However, whenever a deployment is required is random, the deployment quality should be determined and is the challenging issue in WSN. Thus, the qualities of the deployment of sensor nodes can be assessed by applying appropriate measures and determine the weakness in the sensing coverage area of WSNs. This can be determined by calculating the probability of intrusion detection and the success ratio. One of the fundamental issues in WSN and its characteristics requires considering the parameters like sensing range and node density w.r.t high probability of detection. In this section, the model of probabilistic detection of intrusion is discussed which provides a high accuracy rate and thus obtains high energy saving.

4.1. *Probability and its Sensing Model*

A network topology deployed randomly is considered, with all the nodes deployed randomly in the area of interest. The quality of deployment increases as there is an increase in the probability of detection. The sensing model is designed and the probability is carefully coupled and the particular application of sensors. The type of the device used for sensing, and adopt a sensing model to compute the probability whereas the sensor node detects the occasions situated in the region of the sensing region. All the sensor nodes are unspecified that they have a similar sensing range. The node density λ with the number of sensor nodes 'N' are deployed in the sensor network distributed uniformly over the surface area S. For every sensor node, the probability of sensors within the sensing range RSENS is determined at a distance of point by using the Bernoulli experimental rule, where the likelihood of detecting success rate is given by,

$$p = \frac{\pi R_{SENS}^2}{S_{area}} \quad (1)$$

Thus, a binomial distribution is formed within the distance RSENS for the number of sensor nodes.

The probability calculated earlier is further represented by the succeeding equation:

$$P(n = k) = \frac{\left((2R_{SENS}l + \pi R_{SENS}^2)\lambda\right)^2}{k!} e^{-(2R_{SENS}l + \pi R_{SENS}^2)\lambda} \quad (2)$$

The single sensor is used for detection and the probability$\overline{P} = e^{-\lambda S}$, when no sensor can detect an event in the area within the sensing range is considered in the scenario. The complement of the probability $\overline{P}$ is that, at least one node that is a sensor detects an event. Thus, the probability of sensing model is represented by,

$$P(n \geq 1) = 1 - \overline{P} = 1 - e^{-\lambda S} \quad (3)$$

Agreeing to the scenario of intrusion, the starting point of an intruder is taken randomly in the area of sensing. Thus, the moves are made arbitrarily as demonstrated in figure 1. To calculate the probability of an intruder that cannot surpass the value set as the threshold distance l_{THR} is as below

$$P(n \geq 1, 0 \leq l < l_{THR}) = 1 - e^{-(2R_{SENS}l_{THR} + \pi R_{SENS}^2)\lambda} \quad (4)$$

Let us assume that a distance threshold $I_{THR} = 0$ The probability of detecting an intruder in WSN with N number of random nodes distributed in a region A with a length of edge L considering uniform distribution is represented with the equation below. We also consider that all the sensors those are deployed are static, thus immediately detection of an intruder can be obtained by the equation below:

$$P_{sensing}(n_{brs} \geq 1, l_{THR} = 0) \geq P_{threshold\_sensing} \quad (5)$$

Thus,$P_{threshold\_sensing}$ is depicted as the threshold probability. All the sensor nodes with its sensing range are set as follows:

$$R_{SENS} \geq \sqrt{\frac{\ln(1 - P_{threshold\_sensing})}{\lambda\pi}} \quad (6)$$

In the applications of WSN, the required numeral of sensors depends on the coverage quality of the sensors. By the deployment of nodes, the degree k for specified coverage, fork>o, the probability that an intrusion detection P within the threshold of Intrusion distance l_{THR}can be represented by the multi-sensing model as below:

$$P(n \geq k, 0 \leq l < l_{THR}) = 1 - \sum_{i=0}^{k-1} \frac{(S\lambda)^i}{i!} e^{-S\lambda} \quad (7)$$

Here, the surface through which the intruder has swept is denoted by S, with a trajectory l as:

$$S = 2R_{SENS} l_{THR} + \pi R_{SENS}^2 = 2R_{SENS} \int_a^b \sqrt{(f'(x))^2 + (g'(x))^2}\, dx + \pi R_{SENS}^2 \quad (8)$$

The sum of partial densities λ is:

$$\lambda = \sum_{i=1}^{m} \lambda_i = \sum_{i=1}^{m} \frac{n_i}{S_i} \quad (9)$$

As the sensing scale and the no. of nodes per elements are increased, the probability of detecting an intruder in both single sensing model and many-sensing model also increases. Thus, the formulae discussed above depict the optimal sensing range values and the node density to cover the total sensing area.

The distance of an intruder l can be calculated by the parametric curve length using,

$$L = \int_a^b \sqrt{(f'(x))^2 + (g'(x))^2}\, dx \quad (10)$$

The intrusion detection probability where the threshold distance l_{THR} exceeding in WSNs with the density of node λ, range of sensing R_{SENS} and node availability p in the single sensing is determined by

$$P(n \geq 1, 0 \leq l < l_{THR}) = 1 - e^{(2R_{SENS} l_{THR} + \pi R_{SENS}^2) p\lambda} (11)$$

The probability P for detecting an intruder within its distance of intruder with thresholdl$_{THR}$ in the multi-sensing i.e. k-sensing in WSNs of compactness of nodeλ, range of sensingR_{SENS} with the availability of node, p is controlled by

$$P(n \geq 1, 0 \leq l < l_{THR}) = 1 - \sum_{i=0}^{k-1} \frac{(Sp\lambda)^i}{i!} e^{(-Sp\lambda)} \quad (12)$$

$$S = 2R_{SENS} l_{THR} + \pi R_{SENS}^2 \quad (13)$$

4.2. *Working Vacation Scheduling of Sensor nodes*

Thus, the problem is to find the quickest change in the detection of an intruder by finding the minimal observations and builds the Bayesian framework using Markov Decision Process (MDP). This formulation takes the parameters as, cost caused due to delay in detection, false alarms, and cost estimated per unit of the network of sensor nodes. At the time slot k, the node at the fusion center finds the observation vector $X_k^{(M_k)} := \left(X_k^{(1)}, X_k^{(2)}, \dots X_k^{M_k}\right) \in R^{M_k}$and then computes the probability of change in detection, π^k. Thus, the fusion center chooses control among the set of available controls, $A = \{stop,\ \cup_{m\in\{0,1,\dots n\}} (continue, m)\}$. once the control is chosen when control$u_k = (continue, m)$, which means that the process of detection is sustained at the next time slot say k+1 consisting of m number of sensors in the wake state. Grounded on the theory of Bayesian and MDP model, the optimal policy is obtained which determines the following rules:

a) **Stop rule**: In this rule, the sensor node at time slot k, chooses the state, stops, and sleeps for a small interval of time, termed as vacation period of the sensor node. It stops at time slot $T^* = \inf\{k \geq 0 : \pi_k \geq \Gamma\}$, where $\Gamma = [0,1]$ is the maximum, and
b) **Wake rule**: For M_{k+1}, there endures a map represented by $M^* : [0,1] \to Z_+$and the optimal policy is obtained by changing the state of the sensor from small working vacation to the wake state for an optimal number of required sensors in the time slot k+1, given as: $M_{k+1} = M^*(\pi^k)$.

Thus, the sleep/wake state of the sensor node is best suitable when a threshold is applied, and many small working vacations can be performed between the time intervals, based on the probability of detection of an intruder.

4.3. *The Bernoulli working vacation Process*

The sink node selects any of the sensor nodes to begin the working vacation when there is no activity required within its range of sensing and thus the vacation time follows the distribution exponentially with the parameter 'h'. During the vacation period, if any data is being transmitted, then the node works at a very low speed. If any node finds that the service completion is needed or if it detects an abnormal behavior at a low-speed rate, in its vacation period, it switches its state from to the busy state, which means that the vacation is interrupted. If none of the nodes are in the completion of the vacation period, then the sink node either makes the sensor as idle or initiates a new transmission of data, with probability p, a single working vacation changing itself to the regular mode. Otherwise, it may also leave for the next working vacation with probability$q = 1 - p$, i.e. multiple working vacations. Once the vacation period is nearly to the end, and if there is more service needed by the nodes, then the sensor node switches to the normal working mode. During the period of working vacation, the service time determines the most general random variable as S_v, with its distribution function, $S_v(t)$.

The primary objective of the proposed methodology as follows:

- Quality in sensor data collection.
- Improve the network's lifetime.

Our scheme uses a multi-sensing adaptive approach to use the positive points of both cluster and chain-based routing patterns. Redundant data traffic is reduced by correlation, so efficiency increased. PSN (Primary sensing node) uses a probabilistic method while sensing in which nodes with greater residual energy has a higher probability to sense and transmit data so that the energy utilization is equalized on each node. The redundancy or similarity between the sensed data can be found or related by GCA (Gray correlation analysis. Bernoulli identical sampling method is used to find the estimated average value of data characteristics.

In the proposed methodology, all nodes are classified into the following types:

- Primary Sensor Node (PSN)
- Secondary collection node (SCN) / Cluster Head

Primary Sensor Nodes senses the environment and initiates transmission of data to sink.SCN or CH is the second level in the transmission methodology and which collects all data & optimizes the data by detecting duplicate data & tries to increase the accuracy and forwards it to the next intermediate nodes to connect the sink.

4.4. *Algorithms Designed*

Algorithm 1: Cluster head selection and association model

Phase 1: Formation of clusters

Sensor networks strongly alter topology without centralized control to embrace the shift in topology; Adaptable hierarchical cluster at the K-level is used. A k-level clustering chain of authority is primarily helpful in reducing the network's power utilization compared to Low-energy Localized Clustering (LLC). This clustering is especially feasible in reducing the system's contrasting vitality usage (LLC) because it requires short-run transmission into account. It also ensures flexibility in the selection of routing for modifications that affect both the system and the environment. Thus, the X-LLC cluster is used as X-LLC enables the size of the cluster to be reduced by considering the radius by using separate energy concentrations. This delivers a significant benefit in terms of minimizing broadcast power consumption concerning standard cluster-forming hierarchical algorithms [10].

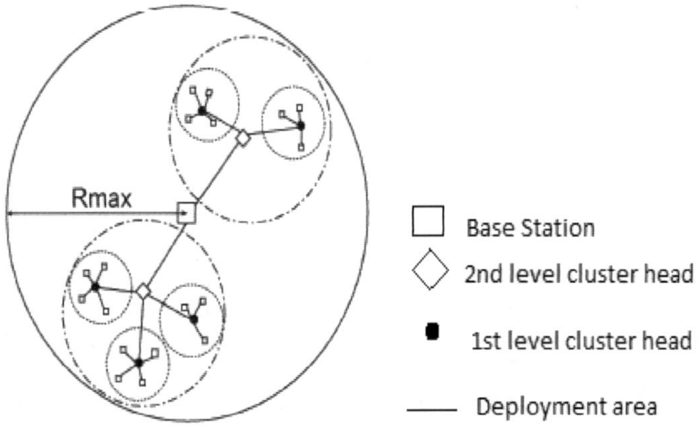

Figure 1: Example of a hierarchical structure (X-LLC, k = 2) [10]

Phase II: Establishment stage

The foundation stage starts after the summit of the political decision technique and ensnares k-explicit affiliation sub organizes that are practiced in the best down manner beginning from the Base Station to straight centers. At this stage, 1stkth-level group heads describe themselves to the BS, which comes back to TDMA table. At that reality, the (k– 1) th level bunch scrambles toward the closest kth - level group head, that answers by giving the TDMA table; the cycle rehashes down to the typical specially appointed hub level

In adding up, the accompanying similarly remains constant.

I. Each cluster head reins over scarce hubs.
II. Simple center points find the nearest bunch head with the withdrawing of a single hop.
III. The program extent of direct center points can be decreased with respect to the one required by LLC. Accordingly, broadcast needs a little measure of intensity and the bury group block diminishes.

Algorithm 2: Energy consumption model:

Every sensor node will have two stages active and sleep. Whereas working status is classified into three stages: working, the transition from active to sleep, and vice versa. The details can be shown as follows

$$E_{sen} = \sum_{j=1}^{m} T_{tra}(B_j)\left(e_{ba}(B_j)\right) + e_{sa}(B_j) + P_a(B_j)T_a(B_j) \quad (14)$$

Where $T_{tra}(B_j)$represents the number of sensor Bj revolved on or turned off, $P_a(B_j)$ and $T_a(B_j)$ are the power and operational time the sensor Bjis in active status, correspondingly. $e_{as}(B_j)$and $e_{sa}(B_j)$are the energy utilization of node state alteration from energetic to sleep and sleep to active of sensor Bj, correspondingly. An energy utilization model of the transmission module is presumed as the same as that in sources.$E_T(j,d)$ and $E_R(j)$are the energy utilization of communicating and obtaining j bits data done a distance d.

$$E_T(j,d) = \left(E_{T-elec} + \varepsilon_{amp} * d^{\acute{\alpha}}\right) * j \quad (15)$$

$$E_R(j) = E_{R-elec} * j \quad (16)$$

where E_{T-elec} and E_{R-elec}*j represents the electronic circuit expenses of the transmitter and receiver, and they are independent with the communicating distance. ε_{amp} [Joule/ (bit/ ma)] denotes the energy costs to send a bit over a distance d with a satisfactory signal to noise ratio and it's constant. $\acute{\alpha}$ represents path loss interpreter which relays on the properties of the broadcasting channel. Generally, it is assumed that

IOP Conf. Series: Materials Science and Engineering 1049 (2021) 012007 doi:10.1088/1757-899X/1049/1/012007

$$E_{T-elec} = E_{R-elec} * j = E_{elec} \quad (17)$$

5. DISCUSSION RESULTS AND NUMERICAL EXPERIMENTS

In the proposed work, we have evaluated the model for the probability of intrusion detection in WSN by using the simulation Software NS2. According to the model probability, the model is evaluated based on the node density, distance of an intrusion, and also the sensing range. We consider N number of static sensor nodes in the field of interest in WSN, which are uniformly distributed and are independent. The results show the probability of intrusion detection P having at smallest one sensor node that discovers in the detecting range RSENS and N number of sensor nodes. As the detecting range and the number of nodes increases, the probability of discovery of intruder also increases. The two possible cases are discussed to show the performance of WSN when intrusion is detected.

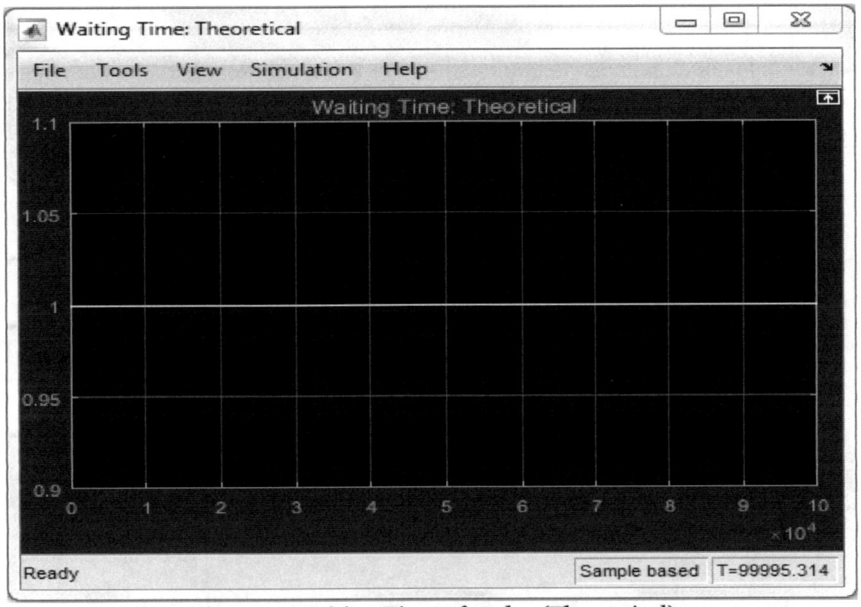

Figure 2: Waiting Time of nodes (Theoretical)

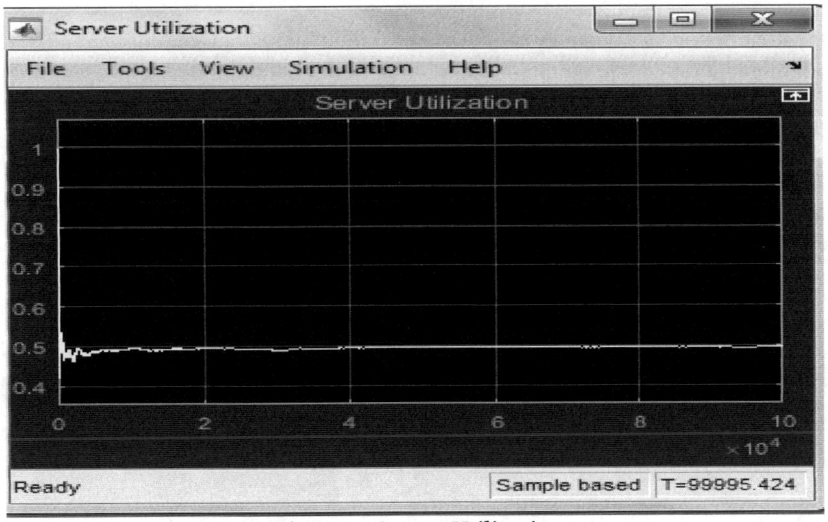

Figure 3: Server Utilization

The proposed approach is compared with EELEACH protocol and EDCSWS and the results are shown as:

IOP Conf. Series: Materials Science and Engineering 1049 (2021) 012007 doi:10.1088/1757-899X/1049/1/012007

1. Dead Nodes:

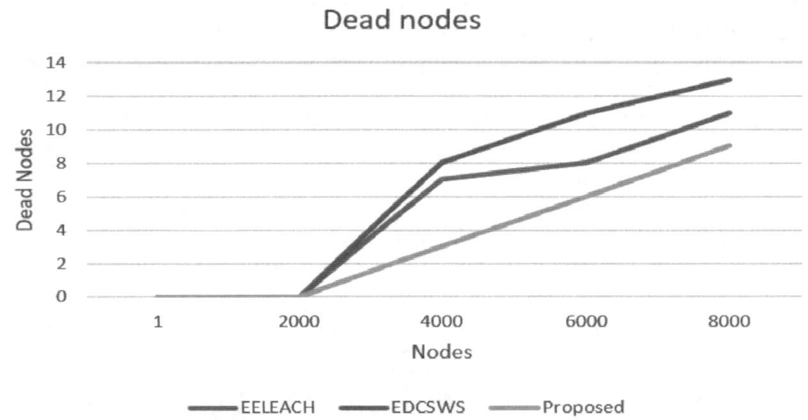

Figure 4: Analysis of dead Nodes

2. Throughput

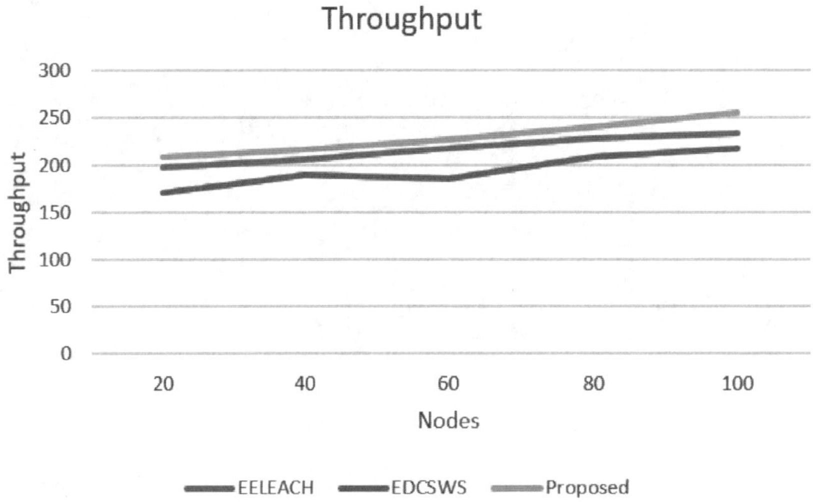

Figure 5: Number of Nodes Versus throughput

This is the number of information bundles conveyed on from a sourcing hub to a target hub for a discrete unit of time. The fig shows the throughput accomplished by both directing conventions. The throughput is recouped as related to other winning strategies. The outcome clearly shows that throughput achieved in our prescribed convention is improved when contrasted with others. With the developing number of hubs, the estimation of throughput is additionally increasing.

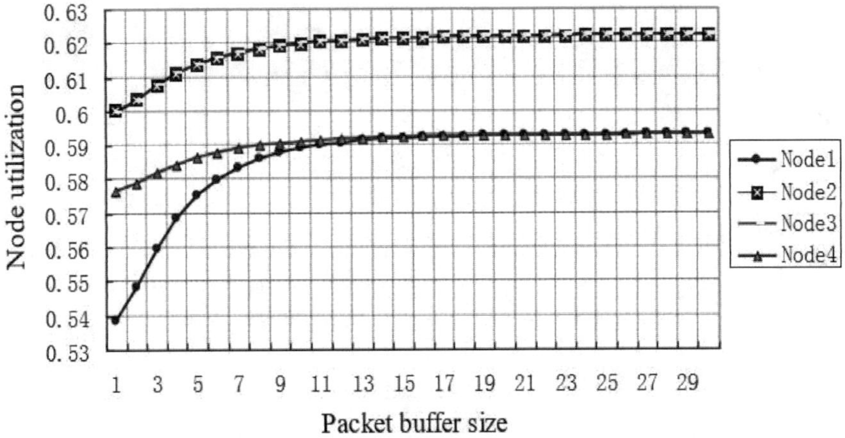

Figure 6: Node utilization v/s packet buffer size

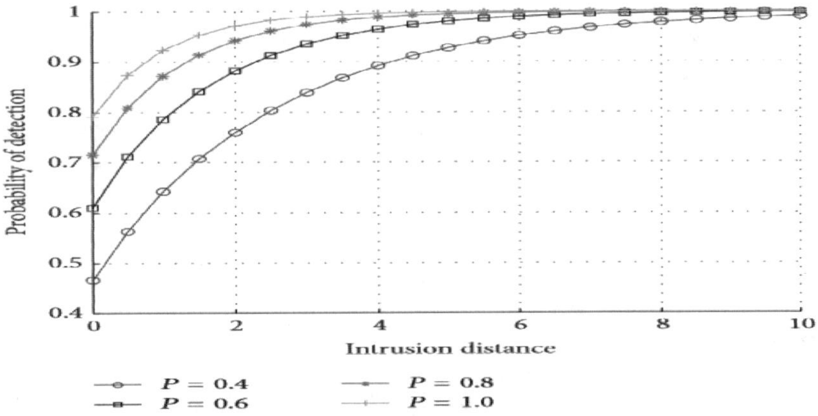

Figure 7: Probability of Detection of Intruder

Figure 7 depicts the curves obtained by the experimental results of the probability of intrusion detection-based model which is considered as a function with intrusion separate for distinct values of rate of availability of nodes, p. If the distance of intruder l increases, the probability of detection P also increases. When the network activity is in normal cycle mode, then the rate of node obtains probability p should be typically less than the value 1.0. It is required that it must be satisfied that the monitored area is to be considered effective without the overall area coverage to be diminished. If the detection of an intruder is successful by a node, a message with an alarm is broadcasted to the entire sensor network. Thus, the detection efficiency can be improved by assuring the effective connectivity of the network. This is illustrated by the rate of node availability p =1.0.

The comparison is performed and tabulated in table 1. The QoS parameters are considered for measuring the performance of the proposed algorithm based on Bernoulli scheduling vacation methodology. Our proposed work is compared with EEEACH and EDCSWS based on throughput, transmission delay, energy variance, sensing probability, probability of detection of an intruder. By the simulation and experimental results performed the proposed methodology BVS has obtained better performance results with minimum energy consumption and high throughput.

Table 1: Comparison of performance of the proposed algorithm with other protocols

	Energy Variance (joules)	Sensing Probability	Throughput	Transmission Delay (ms)	Probability Detection of Intruder
EELEACH	2-8	0.0022	200	0.27	0.4
EDCSWS	2-18	0.025	230	0.31	0.6
Proposed Algorithm (using Bernoulli scheduling)	0-5	0.15	250	0.25	0.8

6. CONCLUSION

The energy-saving is an important factor for the efficient working any wireless sensor networks. Since the sensor nodes are always active for detecting an intruder, the energy consumption becomes high. To reduce the consumption of energy, sleep/wakeup, also called small interval vacations are used for the sensors which are almost considered to be like the mechanism of fluid queues. Here in the proposed work, we are introducing the Bernoulli scheduling rule and the sensor nodes have a vacation state whenever the sensor need not be active, to save the energy consumed. Based on the sensing coverage area, the intruder distance l is computed considering the probability of intrusion detection. The probability obtained shows the better performance of the network with high availability of node and also the working vacations set for every sensor node to save energy and lifetime of the network. The waiting time of the nodes is simulated and shown in the results. Also, server utilization is calculated as discussed. Results are being compared with EELEACH (energy-efficient and EDCSWS and found that our proposed algorithm is more efficient in comparison to the EELEACH and EDCSWS. Thus, our results of the proposed work enable us to propose and analyze the sensor network and help us for considering the diagnostic parameters of a network such that the network requirements are met.

References

[1]. Lasassmeh, S., M., Conrad, J., M.: Time synchronization in wireless sensor networks: A survey. Conference Proceedings-IEEE SoutheastCon 2010: Energizing Our Future, 242-245, Concord. (2010).

[2]. Son, J., H., Lee, J., S., Seo, S., W.: Topological key hierarchy for energy-efficient group key management in wireless sensor networks. Wireless Personal Communications, Vol. 52, No. 2, 359-382. (2010)

[3]. Tan, L.; Tang, S. Energy harvesting wireless sensor node with temporal death: Novel models and analyses. IEEE/ACM Trans. Netw. 2017, 25, 896–909. [CrossRef]

[4]. Tang, S.; Tan, L. Reward rate maximization and optimal transmission policy of EH device with temporal death in EH-WSNs. IEEE Trans. Wirel. Commun. 2017, 16, 1157–1167. [CrossRef].

[5]. Hosen, A.S.M.S.; Cho, G.H. An energy-centric cluster-based routing protocol for wireless sensor networks. Sensors 2018, 18, 1520. [CrossRef] [PubMed].

[6]. Mostafaei, H. Energy-efficient algorithm for reliable routing of wireless sensor networks. IEEE Trans. Ind. Electron. 2019, 66, 5567–5575. [CrossRef].

[7]. Srbinovska, M.; Cundeva-Blajer, M. Optimization methods for energy consumption estimation in wireless sensor networks. J. Sustain. Dev. Energy Water Environ. Syst. 2019,

7, 261–274. [CrossRef].
[8].Kempa, W.M. Buffer overflow duration in a model of WSN node with power-saving mechanism based on SV policy. In Proceedings of the 23rd International Conference on Information and Software Technologies (ICIST), Druskininkai, Lithuania, 12–14 October 2017; pp. 385–394.
[9]. Kempa, W.M. Transient queue-size distribution in a finite-capacity queueing system with server breakdowns and Bernoulli feedback. In Proceedings of the 43rd International Conference on Applications of Mathematics in Engineering and Economics (AMEE), Sozopol, Bulgaria, 7 December 2017.
[10]. J. Jia, X. Wu, J. Chen, and X. Wang, "Exploiting sensor redistribution for eliminating the energy hole problem in mobile sensor networks," EURASIP Journal on Wireless Communications and Networking, vol.2012, article 68, 2012.
[11]. L. Tao, Z. Wang, and Z. Liu, "The GI/M/1 queue with Bernoulli- schedule- controlled vacation and vacation interruption," Applied Mathematical Modeling, vol.37, no.6, pp.3724– 3735, 2013.
[12]. Smith, J. M. G. (2015). Queue decomposition & finite closed queueing network models, Computers & Operations Research, 53, 176-193.
[13]. Diabat, A., Dehghani, E. and Jabbarzadeh, A. (2017). Incorporating location and inventory decisions into a supply chain design problem with uncertain demands and lead times, Journal of Manufacturing Systems, 43, 139–149.
[14]. Zhou, B.; Ahn, D.; Lee, J.; Sun, C., Ahmed, S., Kim, Y. A passive tracking system based on geometric constraints in adaptive wireless sensor networks. Sensors 2018, 18, 3276. [CrossRef] [PubMed].
[15]. Ghosh, S.; Unnikrishnan, S. Reduced power consumption in wireless sensor networks using queue-based approach. In Proceedings of the 5th IEEE International Conference on Advances in Computing, CommunicationandControl (ICAC3), Mumbai, India,1–2December2017; doi:10.1109/ICAC3.2017.8318794. [CrossRef].

IOP Conf. Series: Materials Science and Engineering 1049 (2021) 012008 doi:10.1088/1757-899X/1049/1/012008

An enhanced intelligent classification approach to improve the encryption of big data

Gitanjali Gupta [1] , Kamlesh Lakhwani [2]

[1] *Department of Computer Science and Engineering, LPU*

[2]*Department of Computer Science and Engineering, LPU*

E-mail: [1]gitagupta32@gmail.com
[2]kamlesh.20980@lpu.co.in

Abstract. Progression in cloud applications have been hindered due to critical issues such as Data security and privacy. The aggravating concern remains the ability and access of cloud operators in acquiring sensitive data. Consequently, usage and excitement of cloud computing has remained lukewarm, with users remaining skeptical of the level of confidentiality promised. Principally, the financial industry and governmental agencies have shown lackluster and inimical enthusiasm of these indications. The paper serves to analyze the aforementioned issue, and present a creative proposal into cryptography, as a means to virtually restrict cloud service operator's access to sensitive data. The free-will of cloud operators in handling sensitive data will be undermined. Through this method, file is divided with more accuracy using an intelligent classification technique. Alternatively, a different method can be utilized to find out whether data need splitting to obtain lesser operating time and minimize storage space. The results show that the approach can resolve innumerable risks associated with cloud computing, whilst requiring sufficient computing time using a very good intelligent machine learning classification techniques. As such, novel approach has been proposed which is entitled as a model for Sensitive Encrypted Storage (SES). In this model, three proposed algorithms Convolution Neural Network with Logistic Regression, Elliptic-curve Diffie Hellman-Shifted Adaption Homomorphism Encryption and Decryption have been integrated.

1. Introduction

Cloud computing is the new innovative technology that brings boom to the industry. Earlier when the user is to use some of the features of the application and in any case the local interface is blocked for the usage then one is unable to develop the application at that point of time and entire system of development become to halt [4]. There are many large companies that are efficient in serving cloud interfaces such as Google, Microsoft, Facebook and Amazon and offer their own clouds. The cloud interface is efficient and could be building in minimal investment and one could use the services from any point of time and for much longer time. It helps in migrating data in most convenient way. The customers can opt for flexible services and competitive storage package [1-3]. The other unique services offered by cloud interfaces are IaaS, SaaS and PaaS depending upon the demand of the customers. There exist some security problems in as users are more concerned about the storage of their sensitive data and these are better overcome with the advent of cloud interface. This is due to the fact no knowledge of physical location is being given to the user and due to this there is limited or no chance of access to data and causing breach. This way partnership between cloud service providers and customers is formed. The cloud interface worked well with the big data and provides various solutions as the users' end [5-7].

2. Phases of the proposed model

Below are the different phases that are involved in the proposed model as shown in figure 1:

- Phase 1

In the process covered under this phase, precision, recall and accuracy form the basis for evaluation and analysis. The entire model in proposed phase reflects secure data classification. Protecting the sensitive data is major concern here.

- Phase 2

The process in this phase is responsible for storing the data in the safe environment and also store data that is not of much importance for reliable use. The given phase evaluates encryption approach, time and analyses storage.

- Phase 3

The enhancement in the overall process is the main focus in the phase. This is done to ensure better results in return by making the relevant changes in time, accuracy, parameters, data security and integrity. The success of the process is measured if it is performed in minimal time as well as storage.

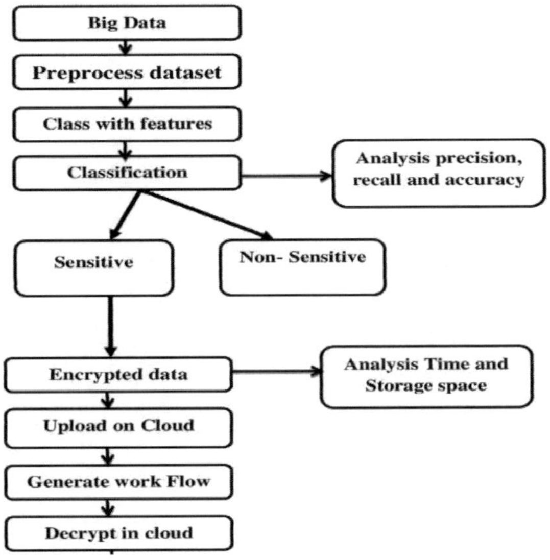

Figure 1. Proposed Methodology

3. Simulation model

The followed algorithm is described below:
The used simulation model mainly worked upon underlined three main algorithms :

- Convolution Neural Network with Logistic Regression (CNN-LR) [6,12]
- The tweet text is taken as the input in the algorithm. After input, the input text is pre- processed with the aim of extracting the main function. Further, in the initial step the mechanism of convolution, pooling, and activation is performed in order to reduce non linearity so that logistic regression could be used. The use of EPOCH enhances the process overall and improves upon the accuracy and all the loss are eliminated.
- Diffie–Hellman-Shifted Adaptation Homomorphism Encryption [8,13]

The above algorithm approach is responsible for performing the data processing before the data is actually transfers or shared against the cloud interface [10,11].

- Elliptic curve Diffie–Hellman-Shifted Adaptation Homomorphism Decryption [9]

The algorithm approach is responsible for performing the data processing before the data is actually transfers or shared against the cloud interface. The alignment of the algorithm is done until it is sent to the cloud processors for actual processing an represents the data in the elliptic curve [14].

4. Results and discussion

The result section is effective in presenting the outputs that are gained from the analysis as well as evaluations that are performed. Initially, the comparison of accuracies of different machine learning techniques is performed. This is achieved by inserting the data that is of same size and further their accuracy rate is calculated by using different classification techniques.

The given figure 2 is representative of different machine learning techniques that follows both accuracy and precision. Again, for the analysis same data size is inserted and is measured for determining encryption time

IOP Conf. Series: Materials Science and Engineering 1049 (2021) 012008 doi:10.1088/1757-899X/1049/1/012008

execution and storage space by figure 3, 4 and 5.

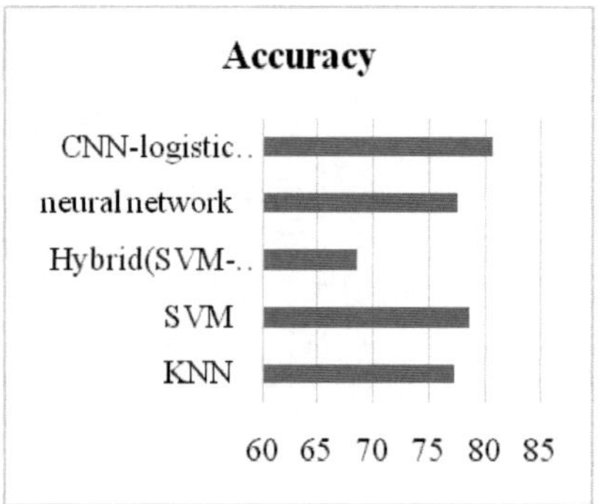

Figure 2: comparison of different classifier accuracy

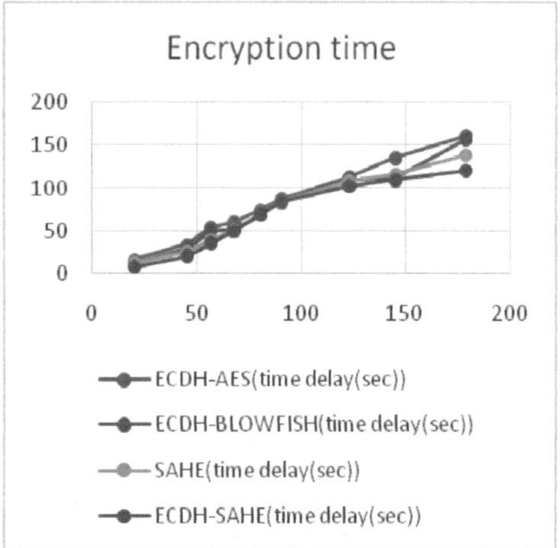

Figure 3. Comparison of various encryption approaches in terms of encryption time

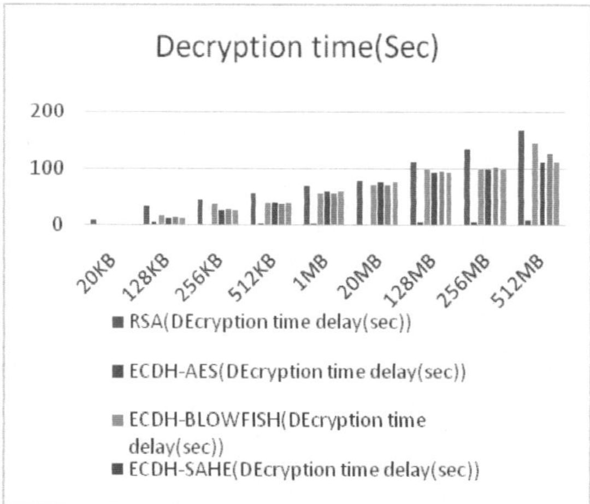

Figure 4. Comparison of various decryption approaches in terms of encryption time

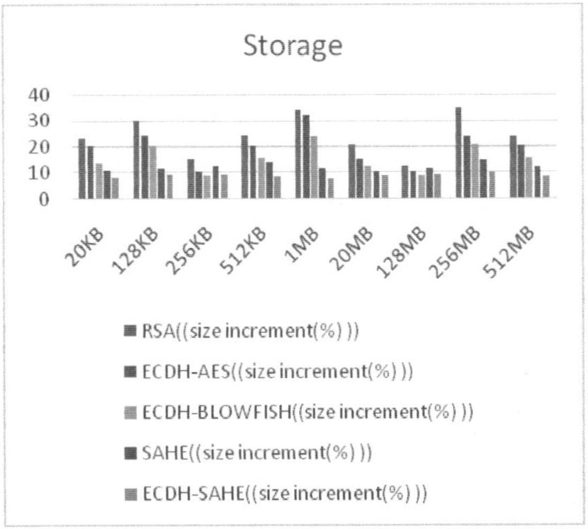

Figure 5. Comparison of various encryp approaches in terms of storage space

4.1 Reason for selecting CNN-LG

- The main reason could be seen as one move from KNN to neural network than the resultant that is achieved is appropriate and selects layered network of convolution.
- There is no scope of non- linearity in the selected model and all features are advances and are linear in nature.
- The classification is further optimized in the form of machine learning.

The experimental scale mostly ranges from 50 MB to 512 MB. The graph obtained as output is reflective of enhancement in the processes of encryption, decryption, time and security. Both the performance of large as well as small data set is calculated and it effectively accommodates. Following are the main reasons that are seen for a change are:

- The binary stream is utilized over unary stream that eliminates overhead time.
- The storage is improved in case of encryption and decryption method that moves in left and right direction respectively.
- Shifting analysis is mainly concerned with binary streaming.

IOP Conf. Series: Materials Science and Engineering 1049 (2021) 012008 doi:10.1088/1757-899X/1049/1/012008

5. Conclusion

To conclude, paper mainly talks about the services and benefits offered by the cloud interface. Further all the algorithms that are used for this purpose are highlighted in the paper. On analysis it is found the proposed methodology is effective in protecting the cloud-based methods against various threats and challenges. Later, the topic and concept related to encryption and decryption are also highlighted into document.

6. References

[1] Sharma S, Garima G and Laxmi P 2014 A survey on cloud security issues and techniques *arXiv: preprintarXiv* **1403**

[2] Grover J, and Sharma M 2014 Cloud computing and its security issues—A review *In Fifth International Conference on Computing Communications and Networking Technologies (ICCCNT)* pp. 1-5

[3] Hashizume K, David R and Eduardo M, and Eduardo F 2013 An analysis of security issues for cloud computing *Journal of internet services and applications* **4**

[4] Almorsy M, Grundy J and Ibrahim S 2011 Collaboration- Based Cloud Computing Security Management Framework *IEEE conference of cloud computing, Washington (DC)* pp. 364-371

[5] Yibin L, Keke G, Longfei Q, Meikang Q and Zhao H 2016 Intelligent Cryptography Approach for Secure Distributed Big Data Storage in Cloud Computing *Information Sciences* **387** (10.1016/j.ins.2016.09.005)

[6] Gitanjali and Lakhwani K 2019 A novel approach of sensitive data classification using convolution neural network and logistic regression *Int. J. Innov. Technol. Explor. Eng.* **8** pp. 2883–2886 2019

[7] Sood S 2012 A combined approaches to ensure data security in cloud computing *ACM Journal of Network and Computer Applications* **35** pp. 1831– 1838

[8] Sengupta N and Holmes J 2013 Designing of Cryptography Based Security System for Cloud Computing *IEEE International Conference on Cloud & Ubiquitous Computing & Emerging Technologies* pp. 52-57

[9] Somani U, Lakhani K and Mundra M 2010 Implementing digital signature with RSA encryption algorithm to enhance the Data Security of cloud in Cloud Computing *In Parallel Distributed and Grid Computing (PDGC) 1st International Conference IEEE*

[10] Rewagad P and Pawar Y 2013 Use of digital signature with Diffie Hellman key exchange and AES encryption algorithm to enhance data security in cloud computing. *Communication Systems and Network Technologies (CSNT) International Conference, IEEE* pp. 437-439

[11] Kumar Y and Kaul S 2019 Effective use of the Machine Learning approaches on different clouds *In Sustainable Computing in Science, Technology and Management, International Conference* pp 892-897

[12] Rozenhaimer M, Li A, Das K and Chirayath V 2020 Cloud detection algorithm for multi-modal satellite imagery using convolutional neural networks *In Remote Sensing of Environment* **237**

[13] Namasudra S 2019 An improved attribute based encryption technique towards the data security in cloud computing *In Concurrency and Computation* **31**

[14] Kapoor N and Kumar Y 2020 The efficient management of renewable energy resources for vanet cloud communication *In nature inspired computing applications in advanced communication networks* pp 228-253

Gazetteer based unsupervised learning approach for location extraction from complaint tweets

[1]Pranali Yenkar
[2]Dr.S D Sawarkar

Datta Meghe College of Engineering Airoli, Navi Mumbai
E-mail - [1]pranaliyenkar@yahoo.com
[2]sudhir_sawarkar@yahoo.com

Abstract: In the era of smart city, social media platform like twitter has become the most powerful medium for sharing urban issues. Extracting the locations from the complaint tweets is very much helpful to local government for monitoring and performing spatial analysis of civic issues at city level. We propose a natural language processing and regular expression based system to identify probable locations of complaints in terms of landmarks (hospital, mall, railway station), street level (e.g. road marg), region (e.g. nagar, colony) and suburb. Our system has used n-gram matching, string similarity and spatial proximity measures to confirm and disambiguate the extracted locations using GeoNames and OSM gazetteer data. Proposed system has achieved the F-score of 93% which is very promising. Locations are also displayed on map for better situational awareness across the city.

1. Introduction

According to the forecasting*, around 65% of the population throughout the world will reside in cities by 2050.Tremendous growth in urbanization leads to many urban issues in terms of safety,comfort,inter city transportation, standard of infrastructure and lots more [1].In contrast to the traditional governance where citizens are needed to report their problems in person and wait cluelessly about the status of it, smart governance promotes the use of information and communication technologies to speed up the whole process. Social media like Twitter is the favored platform to share all the issues and inconvenience caused to the residents due to its vast reachability and high speed [2,3].Along with the type of complaint, location where the issue exists plays vital role in resolving the issues in timely manner. Especially for safety or crime related issues, correct information regarding location is lifesaving.

Geocoordinates information of the tweet indicating the place of its origination [4] may not always be the place of an issue as the user can write the tweet from one place mentioning the issues he faced in another place. Again almost negligible no of users 'geo-tag' the tweet so tweet text becomes an ultimate source of location information. However extracting the location information from tweet is very challenging due to it informal and unstructured format. Named entity recognition, a popular method for location identification of formal data seems be ineffective in case of noisy tweets. Since supervised machine learning techniques needs vast quantity of labeled data, some studies used unsupervised approach to identify probable location names which are confirmed when matched exactly with the gazetteer records. This approach did not always work due to informal writing style of citizens e.g. same address may be written in different ways like 'BD Patil Marg' or 'Bhikaji D patil Marg' which will not matched with the entry 'Bhikaji Damaji Patil Marg' present in gazetteer. Additionally, location disambiguation at the city level like handling street names having same names but different locations is hardly studied in previous studies. It is observed that out of many locations extracted from a tweet which is the best suitable to display on map is not considered in earlier studies.

So purposes of our study are:

1. Identify all possible patterns of location mentions in tweet
2. Explore novel tweet preprocessing steps to locate most probable location mentions

IOP Conf. Series: Materials Science and Engineering 1049 (2021) 012009 doi:10.1088/1757-899X/1049/1/012009

3. Disambiguation of location names at city level
4. Display the exact location on map

So we propose a novel method of automatically identifying the locations from the tweets posted by the users to complain about the facilities within the Mumbai city. Here we are using various NLP techniques in combination with the regular expressions to extract the possible locations from the tweets and verified with the location information source like GeoNames and Open Street Map gazetteer. Several disambiguation rules are applied to handle the location name ambiguities which are showing good results. Locations are also arranged in hierarchical order to represent on the map. Organization of article is as follows: Next section discusses the existing approaches used for location extraction which is followed by detail description of the proposed methodology applied in this study. At last, experimental set up is explained in details along with the result comparison and further concluded with our finding.

2. Related Work

Extracting location information from the text has been highly interesting area for research owing to its promising application in day to day life. Some of the existing approaches used for it are discussed below.

Named Entity Recognition (NER) was employed by [5, 6] for detecting the locations from twitter data. Supervised learning method based on Conditional Random Fields (CRF) [7], maximum entropy model [6] naïve Bayes, Support Vector Machine and Decision Tree[8] are employed by annotating large amount of training data. Publicly available tools like TwitterNLP [5], Stanford NER [9], Google Cloud and OpenNLP** are accessible to extract the location information from content.

Semi supervised approach [10] based on structured perceptron and beam-search has employed on Foursquare data. Gazetteer based methods are also popular due to readily available data sources like GeoNames and Open Street Map data.NLP tools have been used [11] along with the gazetteer for geoparsing the location information from unstructured content. Part of Speech method that identifies the Noun terms was combined with Named Entity Recognition techniques to label them as Person, Location or organization. To handle the location names same as the person names ,pair strength approach is used.Abrol and Muthya[12] integrated various data sources like gazetteer,Wikipedia,traffic data and landmark databases to find the address of the house mentioned in Craigslist advertisements and display it on goggle map. Author has used heuristic-pruning-based algorithm to identify correct location with accuracy of 85% .An unsupervised method using NLP tool is presented [13] to find the location names from the tweet contents. Here authors have used various techniques like parse tree for noun disambiguation, regular expression, dependency graph and NP chunker to find possible place information from the tweet content. Further the extract information is compared and verified against the gazetteer prepared using two different data sources i.e. GeoNames and OSM to filter out the real location names.Auhors have achieved the highest F-score as 81% with GeoNames gazetteer whereas giving the highest precision using OSM gazetteer as 88% but a with poor F1sore of 49% only. An approach based on n- gram and noun phrase identification is proposed [14] to identify locations specified in terms of landmark, street address ,suburb,city,region and country from tweets. The approach is based on use of syntactic parse tree along with the gazetteer created from GeoNames database information. Here the n-grams exactly matched with the gazetteer entries are considered as locations but if the same locations are mentioned in different pattern unlike present in gazetteer then the locations will not be matched and considered as invalid e.g. If entry in gazetteer is 'Fountain Hotel' and user mentioned as 'Hotel Fountain' then it will not consider it as valid location. A language model has been created[15] based on probability distribution of words using the combination of data sources like OSM,GeoNames data and DBpedia.Here gazetteer entries are converted into unigram, bigram and trigrams to obtain all possible valid location terms. Augmenting and filtration techniques applied on it further improved the result. Author [16] has empathized on importance of predicting weather the tweet contains locations or not for highly précised location extraction task. Out of the various features used to train the machine learning model which predict present of location mention in tweet, feature indicating presence of term in gazetteer and feature indicating preposition before the proper noun were the most important features.

3. Proposed Methodology

This section elaborates each step of the proposed system as mentioned in figure 1.

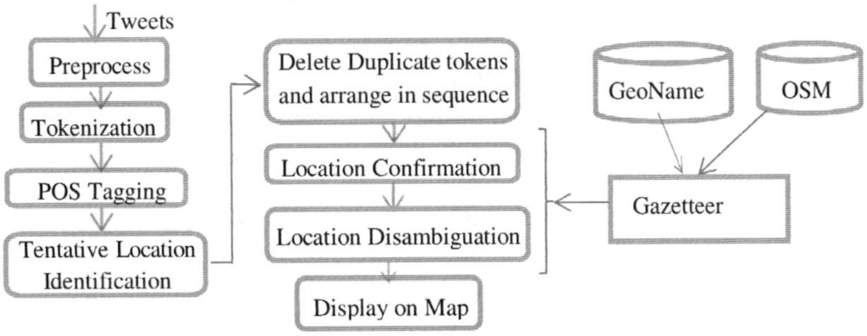

Figure 1: Steps in proposed system

3.1 Preprocess:

Tweets are unstructured in nature so generally consist of lots of unwanted information which could affect the performance of location identification. Eliminating the noise from tweets is called as tweet preprocessing which includes different steps as mentioned below. Here several dictionaries are prepared manually for taking care of spell correction, slang words, abbreviations and stop words removal .Even though case folding is the preferred preprocessing step, in this scenario it is avoided to retain the proper nouns as most of the location names are present in terms of proper nouns.

Steps:

1. Separate Camel Case words like e.g. 'RoadsOfMumbai' to 'Roads of Mumbai'
2. Separate joined alphanumeric and digit terms e.g. 'Chembur74' to 'Chembur 741'.
3. Eliminate @, #, links, email, extra spaces and replace repetitive word with correct word
4. Spell Correction and Abbreviation expansion like rd to road
5. Remove stop words and lemmatize the words

Additional preprocessing technique like separating the concatenated words like 'pushpapark' to 'pushpa park' or 'anushaktinagar' to 'anushakti nagar' is proposed here to improve the accuracy of the location identification.

3.2 Tokenization and POS tagging:

After preprocessing, the cleaned text is converted to series of tokens and part of speech tag for each of the token is identified using NLTK POS.Part of speech plays vital role in identifying the tentative location terms as most of the locations are mentioned in the form of nouns. Along with nouns ,sometimes numbers are also part of the address like '90 feet road Mulund',so we need to extract all the tokens having POS tags as variations of nouns like singular noun, proper noun etc. along with preposition of places and numbers.

3.3 *Probable Location identification:*

Identifying the exact place in the tweet where the location information can be found is not a simple task owing to the unstructured nature of the tweet contents. Proper nouns, place prepositions and landmarks plays vital role in predicting the location in the tweets. Some of the very common place prepositions and landmarks are as mentioned in table 1. After manually inspecting several tweets, various patterns of mentioning the location names are observed which are described as follows.

1) Hash tag and mentions: Hash tags and mentions proved to be an important source for location identification when any other location information is not present .e.g. @MumbaiPolice or #mumbai

2) 'Location/Address'+NN/NNP/CD: location may be mentioned after 'Location' or 'Address' keyword.

3) NNP/NN/PP+Dir: Most of the times the locations are present in form of sub region like 'Mahim east','Andheri West' so direction words are indicative of the locations names.

4) PP+2(NNP/NNS/NN/JJ) : Most probable place to find the location information is after the preposition of place so two words are extracted after every preposition of place present in the text

5) PP+ (NNP/NNS/NN/IN/JJ/CD)+ Landmark :Even though place prepositions are best indicative of the

place names, for detailed location information, words between preposition and landmarks are extracted. 6)2(NNP/NNS/NN/JJ)+Landmark:Sometimes the tweet starts with location names followed by landmark so two words before the landmark words as mentioned in list are extracted as tentative locations

3.4 Elimination of Preposition, Duplicates and Rearranging:

For every preprocessed tweet, once all the regular expression are applied, results are merged together to form a list of tokens. Result may also containg the preposition of places like 'on',' near' which are not part of location names so we are eliminating these words from the result obtained. Along with the tokens, their position is also retained. There are chances that the duplicate tokens will be present in the output list so all the duplicate tokens are deleted first and then rearranged in the same sequence in which they are present in the cleaned tweet using the token positions. Arranging the tokens in the proper sequence improves the chances of identifying the correct location name from the gazetteer.

Table 1: Example of prepositions, landmarks, street level and direction words.

Domain	Words
Place Prepositions	above,across,after,against,along,among,around,at,behind,below,beside,between,by,inside,in, into,near,nextto,off,on,onto,opposite,outof,outside,to,towards,of,close to, down,from,in front of
Point of interest (Landmark)	park,chs,building,chawl,colony,office,market,mall,underpass,workshop,estate,hotel,hospital, school,college,restaurant,junction,lake,river,hill,mountain,bridge,flyover,cinema,temple, chowk,naka,station,depot
Street level	road,marg,street,lane,nagar,highway
Direction	East, west

Table 2: Probable patterns of presence of locations in tweets.

Type	Pattern	Example
Hash tag	#sign followed by Proper noun	Pot holes after pot holes. Surface of the moon right here at **#bandra** #Potholes **#Mumbai**
Mention	@ sign followed by Proper noun	**@Mumbai**Police Today I'm still standing at bus and no bus. I should get late today also?
'Location/Address'+ NN/NNP/CD	Location or Address words followed by Proper Noun ,Nouns or digits	I Feel Betrayed by the government while Travelling on such Injurious prone Pothole Filled Roads @mybmc @MumbaiPolice Request Your urgent Attention **Location - Sj Marg, Dhanraj mill compound, Lower Parel.**
NNP/NN/JJ + Dir	All the Proper nouns ,Nouns adjectives present before Direction words like East ,west	@MumbaiPolice what is the time limit for street processions? It's past 10:40 pm and there's still loud beating drums on lane opposite Toyota showroom **Malad West.**
PP+ 2(NNP/NNS/NN/JJ)	2 words after prepositions which are either Proper nouns ,Singular noun, Nouns or adjectives	@MumbaiPolice Kindly arrange more male and female personal for security **at lalbaug**,anytime stampede like situation can happen
PP+ (NNP/NNS/NN/IN/JJ/ CD)+Landmark	All the Proper nouns ,Nouns, adjectives, preposition conjunction, digits present between prepositions and Landmarks	Slow movement of vehicles **on Goregaon Jawahar Nagar road towards S V Road** at 6.20 pm. #DahiHandi celebration. @RoadsOfMumbai
2(NNP/NNS/NN/JJ)+ Landmark	2words before landmark which are either Proper nouns ,Nouns or adjectives	**Sion Hospital signal**. Tankers n trucks way ahead of signal n crossing. Traffic jam. @CPMumbaiPolice

Table 3: Probable location string identification example.

Tweet	Preprocessed tweet	Output after applying all Regular Expression	After Deleting Duplicate and rearranging	Final string ready to compare with gazetteer
There is road closure on the Western express high way. Some kind of protest going on right on the Cross road near fountain hotel. Please do not protest like this which can trouble everyone @ZeeNews @MumbaiPolice	road closure on Western express high way kind of protest on on Cross road near fountain hotel Please do not protest trouble Mumbai	[[3, 'Western'],[4,'express'], [5, 'high'],[6, 'way'], [8, 'of'],[9, 'protest'], [10, 'on'], [11,'on'], [12,'Cross'],[13,'road'], [14,'near'],[15,'fountain'], [16, 'hotel'],[3,'Western'], [9, 'protest'], [12,'Cross'], [15, 'fountain'], [12,'Cross'], [15, 'fountain'], [13,'road'], [16, 'hotel']]	[[3, 'Western'], [4, 'express'], [5,'high'], [6, 'way'], [9, 'protest'], [12, 'Cross'], [13, 'road'], [15,'fountain'], [16, 'hotel']]	western express high way protest cross road fountain hotel

3.5 Location confirmation using Gazetteer:

The final string may contain zero, one or more than one location names makes it a challenging task to extract precise location information. Here we have extended further the n-gram approach mentioned in [14] with modifications in order to get more accurate location names. Each of these n-grams is compared with the gazetteer entries for location information. Speed of the location verification has been improved by discarding n-grams containing only words which are part of landmark list like 'road','park','sector' or direction list like 'east', 'west' etc. having high probability to find in huge number of gazetteer entries resulting in noise.

Algorithm1: Location confirmation steps

```
Input   :  String of tentative location (Ip), GeoNames and OSM database (Gzt), LandmarkList, DirWordList
Output  :  List of locations with latitude and longitude (Final_Op)

If Gzt [loc] ==Ip
   Return latitude and longitude of matched entry
   Exit
Else
   N=length(Ip)-1
   While N>0
     ng=find_ngram (Ip,N)          //  n-gram of Ip of length N(e.g. 5-gram,4-gram,trigram etc.)
       For each entry x in ng
            If Gzt[loc]== x
               Op1=Op1+loc, latitude and longitude values ox x
   N=N-1
   Ip1=Ip-All locations in Op1
   If length(Ip1)!=0
       U<-Find all unigram of Ip1
       U<-Remove unigrams from U which are part of LandmarkList or DirWordList
       For each entry y in U
            Find all Gzt[loc] contains y
            Op2=Op2+ loc, latitude and longitude values of y
       For each entry z in Op2
            C<-No of words of Ip present in z
            P<-C/total no of words in z
            If P>74
                 Op3=Op3+loc,latitude and longitude values of z
       Final_Op=Op1+Op3
   Else
        Final_Op=Op1
```

While comparing the n-grams, one important step is to expand the abbreviation like ‘B D Patil marg’ or “TH Kataria Marg” when no matching entry is found in gazetteer entries. Most of the times people use abbreviations of the location according to their convenience like ‘BD Patil marg’ or ‘Bhikaji D Patil Marg’ instead of the correct name i.e. ‘Bhikaji Damaji Patil Marg’ which reduces the possibility of identifying the location name from tweet even if it is present. So whenever 4-gram or 3- gram string containing landmark not matches with the gazetteer entries then gazetteer entries are extracted which contains last two tokens. First letter of the remaining tokens are compared and matched entry will be replaced as expanded location.
It is observed that even if most of the locations are identified by approach used in [14] ,because of the unformatted way of writing tweets, some locations are missed. So here in addition to finding exact match of all n-grams of length N-1 to 1, all the entries from the gazetteer containing unmatched unigrams are also examined in order to extract the initially unidentified locations using algorithm 1.

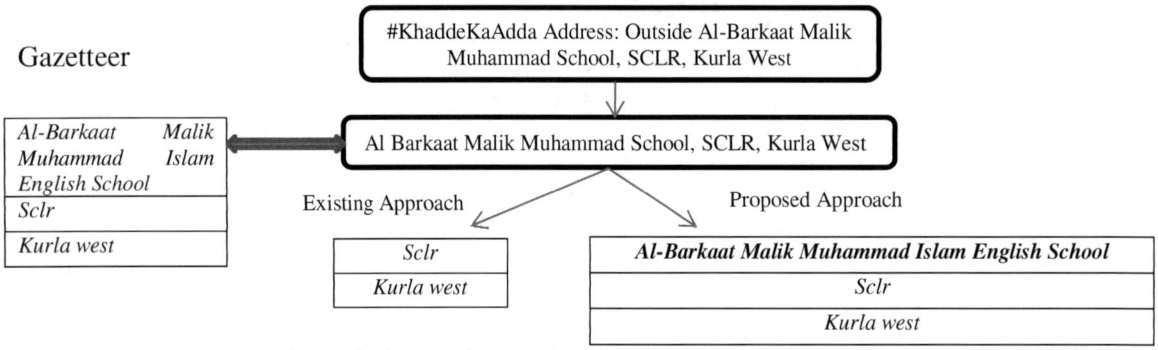

Figure 2: Comparing result with existing and proposed approach

3.6 *Location disambiguation:*
Once location names are verified and finalized using the gazetteer information, location disambiguation is the important task. Disambiguating the location means finding out the most relevant location out of the similar location names.eg. gazetteer has few entries with the same name but different geo- coordinates like ‘S V Road’ which exist in many suburbs of Mumbai like Andheri, Malad ,Bhayandar etc. ,So all the entries will have same names but different latitude and longitude values. So in order to resolve the ambiguity, distance between the ambiguous and unambiguous locations are derived and only the locations within 5 km of radius are selected as valid locations

Algorithm 2: Exact location identification
Input: List of location returned in Algorithm 1(Par_Loc)
Output: Best location information to display on map
1. If count(Par_Loc)==1
2. Return Loc, Latitude and longitude values from Par_Loc
3. Else
4. Find the distance between each location and every other location in the list by comparing their latitudes and longitudes
5. Loc1<-all the locations having distance between them <= 5 Km
6. If Loc1 is null
7. Fin_Loc<-Par_Loc
8. Else
9. Fin_Loc<-Loc1
10. Return location,latitude and longitude information of Fin_Loc

3.7 *Display on Map:*
Final and important step is to display the extracted locations of the map for the better visual understanding of the overall situation. For each tweet, final locations derived using Algorithm 2 may contains multiple locations like 'sion trombay road','trombay','narmadeshwar bus stop'.So the locations are arranged in hierarchical order and first location is displayed on map.

Algorithm 3: Finalize location to display on Map
Input: List of locations returned from Algorithm2 (Fin_Loc)
Output: Geolocating accurate location on map
1. If count(Fin_Loc)==1
2. Display on Map
3. Else
4. Fin_Loc_Hierarchy<-Arrange the available locations in Fin_Loc in Hierarchical Order of Landmark, Marg/Nagar,Highway,region with dir,region
5. Display on map latitude and longitude information of First entry in Fin_Loc_Hierarchy

4. Experimentation and Result
Tweets are extracted from Twitter platform by giving user mentions like @MumbaiPolice, @BMC or hash tags like #Potholes as keywords and classified manually as complaints and non-complaints. Only the tweets mentioning the issues in the Mumbai city and written in English language are considered for study. To illustrate the performance of content based location identifier, random 120 tweets are selected out of which 100 tweets contain at least one location names and 20 tweets do not contain any location reference. Tweets without location mentions are hereby help us to identify any false negative cases. So all the result analysis is performed on these 120 manually annotated tweets.
Baseline algorithm set up is follows:
Baseline-I: Bigrams and unigrams of the possible location text compared with each of gazetteer entry
Baseline-II: Possible location text is converted from n-gram till unigram and *exact matched* with the gazetteer entries. Whenever n-gram matches that n-gram is removed from the cleaned text and remaining text is converted to next level n- gram.
Our approach effectiveness is tested in 2 variations:
In first approach, Algorithm 1,2 and 3 as mentioned above are applied on the tweets sequentially without the novel preprocessing and expandation techniques and approach 2 includes it along with the algorithms to verify improvement in the result. Result clearly shows improvement in the F1-score by using our proposed approaches as compared to the baseline approaches.

Table 4: Result comparison

Method	precision	recall	f1-score
Baseline-I	0.98	0.64	0.78
Baseline-II	0.99	0.73	0.84
Prop App-I	0.95	0.84	0.89
Prop App-II	0.98	0.88	0.93

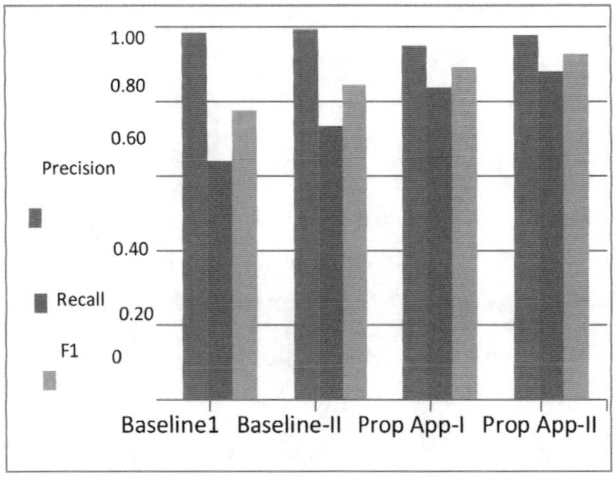

Figure 3: Result Comparison

5. Conclusion

In this study, we put forward an unsupervised approach to identify the locations mentioned in the tweets complaining about the Mumbai city infrastructure. Considering the users informal way of writing, our system extended exact n-gram matching method further and apply string similarity measures up to some optimum threshold value to find probable locations. Here we also presented a novel approach of using the spatial distance between locations to disambiguate a location names at city level. The application of additional preprocessing techniques along with location abbreviation expansion has proven a major factor for improving the overall performance of the system.
.

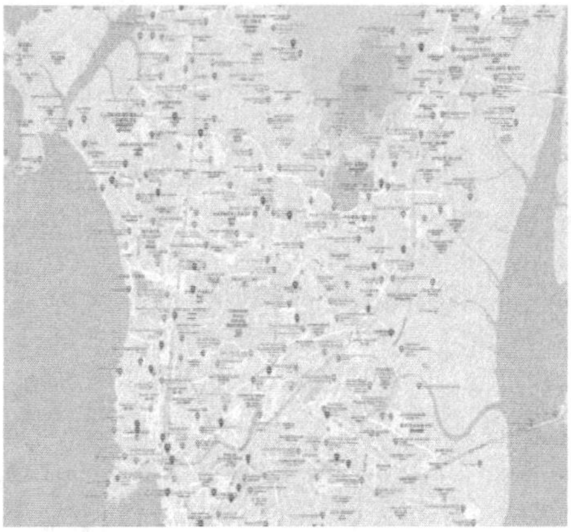

Figure 4: Extracted locations on map

References

* www.un.org ** OpenNLP. http://opennlp.apache.org.

[1] Batty M, 2013 Big data, smart cities and city planning. *Dialogues in human geography*, *3*(3), pp.274-279.

[2] Criado J I, Sandoval-Almazan R and Gil-Garcia J R. 2013 Government innovation through social media 319-326.

[3] López-Ornelas E, Abascal-Mena R and Zepeda-Hernández S, 2017. Social media participation in urban planning: A new way to interact and take decisions *Int. Arch. of the Photogrammetry, Remote Sensing & Spatial Information Sciences, 42.*

[4] Eshleman R and Yang H 2014. Hey# 311, Come Clean My Street!: A Spatio-temporal Sentiment Analysis of Twitter Data and 311 Civil Complaints *Fourth Int. Conf. on Big Data and Cloud Computing* pp. 477-84 IEEE.

[5] Ritter A, Clark S and Etzioni O 2011 Named entity recognition in tweets: an experimental study *Pro. Of conf. on empirical methods in natural language processing* pp. 1524-34

[6] Lingad J, Karimi S and Yin J 2013 Location extraction from disaster-related microblogs *Proc. Int. Conf. on World Wide Web* pp. 1017-20.

[7] Gelernter J and Zhang W 2013. Cross-lingual geo-parsing for non-structured data *Proc.Workshop on Geographic Information Retrieval* pp. 64-71

[8] Indira K, Brumancia E, Kumar P S and Reddy S P T 2019 Location prediction on Twitter using machine learning Techniques. *Int. Conf. on Trends in Electronics and Informatics (ICOEI)* pp. 700-03 IEEE.

[9] Finkel J R, Grenager T and Manning C D 2005 Incorporating non-local information into information extraction systems by gibbs sampling *Proc. Annual Meeting of the Association for Computational Linguistics (ACL'05)* pp. 363-70

[10] Ji Z, Sun A, Cong G and Han J 2016 Joint recognition and linking of fine-grained locations from tweets *Proc.Int.Conf. on world wide web* pp. 1271-81

[11] Lieberman M D, Samet H, Sankaranarayanan J and Sperling J 2007 STEWARD: Architecture of a Spatio-textual search engine *Proc.of the 15th annual ACM international symposium on Advances in geographic information systems* pp. 1-8.

[12] Abrol S, Khan L and Bin Muhaya F T 2013 MapIt: a case study for location driven knowledge discovery and mining *Int. J. of Data Mining, Modelling and Management* 57-75.

[13] Dutt R, Hiware K, Ghosh A and Bhaskaran R 2018 Savitr: A system for real-time location extraction from microblogs during emergencies *Proc Web Conference* pp. 1643-49.

[14] Malmasi S and Dras M 2015 Location mention detection in tweets and microblogs *Conf. of the Pacific Association for Computational Linguistics* pp. 123- 134

[15] Al-Olimat H S, Thirunarayan K, Shalin V and Sheth A 2017 Location name extraction from targeted text streams using gazetteer-based statistical language models. *arXiv preprint arXiv:1708.03105.*

[16] Hoang T B N, Moriceau V and Mothe J 2017 Predicting locations in tweets *Int. Conf. on Intelligent Text Processing and Computational Linguistics*

IOP Conf. Series: Materials Science and Engineering 1049 (2021) 012010 doi:10.1088/1757-899X/1049/1/012010

Optimizing process of Healthcare Assistance at Road Side using Particle Swarm optimization and Person Accident Behaviour

Amreen Khan[1], Dr. M.M. Raghuwanshi[2], Dr. Prashant Borkar[3]

[1]Ph. D Scholar at G.H. Raisoni College of Engineering, Nagpur.

[2]Professor, Department of Computer Engineering, GHRCEM, Pune, Supervisor at G.H. Raisoni College of Engineering, Nagpur.

[3]Head, Department of Computer Science & Engineering, G.H. Raisoni College of Engineering, Nagpur.

amreenkhan786@gmail.com

Abstract. In the worldwide road accidents are very common and happened throughout the day. It is not restricted to time, density of population, geographical location, tropical geography, etc. Due to rapid increase in population and many other factors like driving and safety rules, driving pattern, etc. , the worldwide rate of roadside accidents are also increasing. The African countries is having highest road accident inhabitants along with highest number of fatalities, whereas the Asian countries are the second rank in the world incurred rate of accidents with more fatalities during the road accidents. Hence, In order to handle these road accidents the major concern is to provide quick and fast emergency and health-care services like deployment of quick response team(QrT), first-aid, immediate hospitalization, and various assistance services that helps in reducing the fatalities during the road accidents. Now there is a need to develop effective and less parameterized approach which will helps in reducing the road accidents and provide quick response assistance to the needy. In order to develop the effective and comprehensive solution to the above mentioned problem, optimization is the key ingredient, which results in the evolvement of one of the most popular and effective algorithm called particle swarm optimization. In this paper we are going to evaluate, analyse and simulate particle swarm optimization (PSO) algorithm in order to design and develop novel solution to the above mentioned problem based on accident pattern and the behaviour of a person who met with the road accident.

1. Introduction

Road accidents are one of the incidents where the life of a person is always in trouble. It is not a generic activity, instead it is occurred due to many reasons like violation of traffic rules, road side conditions, driver mental conditions, etc. Many of the cases where the road accidents happened, there is always a need of emergency health care facility. Quick and immediate availability of this facility leads to the safety of persons involved in the accident. There are many requirements of healthcare facilities like first-aid, immediate hospitalization, etc. hence by providing all these facilities on-time we can reduce the rate of fatalities caused in the road accidents. Every government in the countries taking many active efforts in order to provide healthcare facilities to the persons involved in the road accident. Providing the healthcare facilities at the roadside in not new, since all the concern organizations are continuously working on it. Hence if we analyze all the requirements and data collected when the accident happened and after a while, we can make a better system which can serve to provide quick and effective healthcare facilities and assistance to the needy one. Optimizing the complete process of this healthcare assistance, in order to meet the need of availability of resources to the persons is very essential. Now, there is a need to identify the mechanism and technique that will help us to analyze available past experience, transform it into meaningful data and use this data to make an optimize solution to the above discussed problem.

Optimization is a process of selection of one the best solution out of the available list of solutions available to solve a given problem. In most of the cases this may not be an exact solution but it is one of the most deserving and key candidate that will be most suitable as a solution to a problem. While selecting one of the best solution out of available, it is very important to arrange the solution in order of maxima and minima. Optimization techniques are generally classified as: Collaborative in nature like population technique ; biological and evolutionary like evolutionary algorithm; Inspired by swarm knowledge like swarm intelligence algorithm; Inspired by nature phenomenon like nature inspired; knowledge from tight bound processes like Chemical algorithm; inspired by fundamental activities like physical algorithms.

Particle Swarm optimization is one of the most popular technique that uses group based features and knowledge together in order to generate inference as one of the solution to a problem face by a same group. Due to this feature of PSO we can use it as one of the technique that can be implemented along with the data and knowledge involve in the process of providing the healthcare services to the needy one. PSO can be integrated with the useful information, which is generated and available from the road accident, that can be used to distribute the effort and make a collective work on a above problem. Providing a healthcare assistance at roadside is distributive in nature, hence the solution of this problem must be distributive and collective in nature. So, we are proposing a modified particle swarm optimization algorithm based on person's accident behavior that will help our swarm particle to better travel in the direction towards the target. In the next two sections we will present the brief idea about various dimensions of road accident along with PSO and general idea about particle swarm optimization. In the subsequent sections we will elaborate our proposed PSO based on person's accident behavior algorithm followed by evaluation of the above algorithm with the help of sample subset. In last two sections, discussed our findings based on proposed algorithm along with concluding remarks.

2. Road Accidents and Particle Swarm Optimization (PSO)

In the Indian subcontinent road accidents is one of the major cause of concern. In the year 2018, there were 1.5 lakh fatalities in India due to road accidents. Every year around 3-5 % of GDP involves in handling the road accidents. Now, a day's the population of the countries goes on increasing in proportion the sale of vehicles is also increasing. The rate of vehicles on the road goes on increasing day-by-day that makes heavy traffic flow on the roadside, further that heavy traffic flow increases probability of road accidents.

By referring various online survey reports we have conclude that:

1. Two wheelers are more involved in the road accidents rather than the four wheeler.
2. Most of the road accidents are happened at the T-junctions on the road.
3. Mostly the accidents occurred when the driver ignores the traffic and safety rules, ignores the specified speed limit in the various parts of the road.
4. The highways have share of 25% of road accidents rather than the city or inner roads.
5. Roughly 17 deaths happening every hour in India due to road accidents.
6. Road accidents happens more in urban areas than rural areas.
7. Around 65% of the road accidents involvement of youth in India.
8. The rate of involvement in the road accident and fatalities of men is more than the woman across the world.

According to the government officials there are two major factors of road accident:

1. Cause of accident.
2. Handling road accidents.

According to the research community [8] [18] [22] the major cause of concern is:

1. Type of road accident.
2. Cause of accident

IOP Conf. Series: Materials Science and Engineering 1049 (2021) 012010 doi:10.1088/1757-899X/1049/1/012010

3. Type of vehicle involved in road accident
4. Fatality rate
5. Location of road accident, etc.

Chunlu, Wangin his paper [8] proposed (MOPSO) technique to recognize the contributing factors that effects accident severity. Bird s. and co. [9] describes specification based particle swarm optimization algorithm for the improvement in the local convergence. Yare Y and co. [16], presented MD-PSO which is designed for generator maintenance scheduling problem. Engelbrecht A. P [19], illustrated the Heterogeneous H-PSO optimization to solve a multidimensional optimization problem. Phung, M.D and co. [24], shows how optimized path planning can be implemented using Enhanced discrete particle swarm optimization algorithm. Xia, X and co. [25], designed and developed a hybrid optimizer to solve an optimization problem based on firefly and PSO algorithm

In biology, swarm is a phenomenon of knowledge extraction by observation. Many researchers used this concept for developing the solutions for the optimization problem. Through these studies and discussions we have concluded that, if we integrate the PSO along with the information involved in the road accident, it can provide the better solution in order to handle the road accident and will reduce the fatality rate.

Our major objective is to design and develop a novel solution to improve the availability of healthcare assistance services while handling the road accident as well as to contribute towards improvement in the featured solutions available for handling the road side accidents. The solution of the mentioned problem needs to have two basic features: optimization and evolutionary in nature. Due to need of incorporation of these two features we have proposed particle swarm optimization (PSO) which is evolutionary and also has swarm intelligence.

3. Particle Swarm Optimization

The particle swam optimization is a population based algorithm which basically work on the group rather than on the single sample (Fig. 1). PSO is generally applied to the group of random particles and then search optima by updating the generation of every iteration. In each iteration PSO will calculate two best values as:

a. Pbest (Personal best): It is the first solution that meets the criteria to solve a problem. Some time's it is also called as fitness function / fitness value.

b. Gbest (Group best): It is the value that is evaluated by the PSO as a best value, one can have in a group of particles in the network at a given moment of time that meets the criteria of being as a one of the solution to the problem.

After the calculation of the above two base parameters the next is to calculate the next position of the particle by updating the velocity and the present position.

List (v) = list (v) + c1* r * (list (Pbest) – list (present)) + c2* * (list (Gbest) – list (present))
List (present) = list (present) + list (v)
Where,

List (v) ← particle velocity
Present ← current particle

Pbest & Gbest ← Personal best and Group best

c1 & c2 ← learning factor (c1 == c2 && c1←c2←2)
r ← random number between 0 to 1

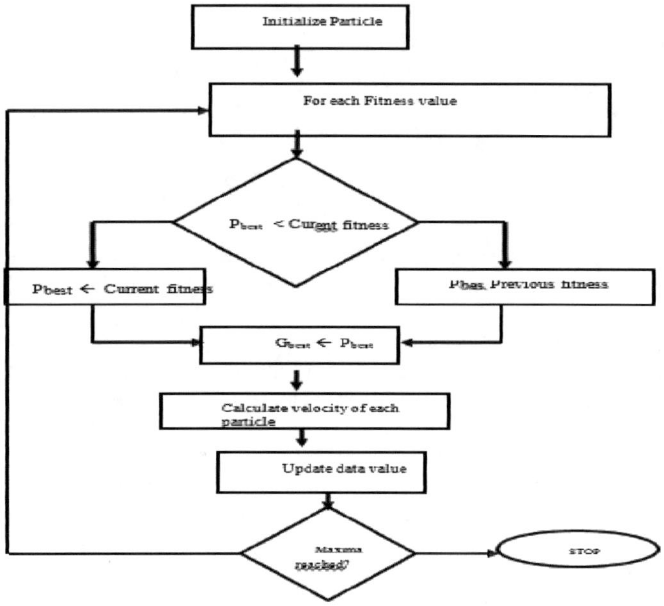

Figure 1. Particle swarm optimization algorithm

4. Proposed Technique and Methodology PSO inspired by Person Accident Behaviour

4.1. Problem statement:

There are list of persons involved in the road accidents, based on the injury they need/referred to healthcare. We have limited number of healthcare resources which can be allocated to a needy person who met with a road accident. Now we need to write an effective and optimized algorithm that make the participation of needy person and the healthcare assistance and facilities to a maxima and helps to improve availability of healthcare services to the person needed.

4.2. Description: The algorithm to solve a problem need some information like how many number of persons met with an accident; criticality of person; availability of resources and etc. using this available information we need to design and develop an optimized solution in order to serve all the needy persons, save the life as much as possible by providing quick and easy access to life care facilities available nearby.

Inputs parameters:

9. NP ← Count of people meet in the accident
10. Upi ← Urgency of saving factor
11. Nv ← No of nearby available vehicles
12. Pvi ←Participation of each Vehicle
13. Ovi ← Occupancy Rate
14. Svi ←Each Vehicle seat count.
7. £ ← urgency of saving people (NP)

Algorithm: PAB-PSO (Person Accident Behavior based PSO):

i. Supi ← sort(descend(£))
ii. Evi ← Cal. (rate of (Ø Vehicle)) {Evi | 1-Ovi}
iii. ESvi ← Evi * Svi
iv. DESvi ← sort(descend(ESvi))
v. For vehicle(V) in {Pvi←1 & Pvi € list(DESvi)}
if Pvi = = 1 : accommodate needy from list(SUpi) & assign to assistance service

=> Gbest ::

Pvi ! = 1 : empty(vehicle)
else:

PSvi ← DESvi * Pvi
DSvi←Sort(descend(PSvi

vi. Pbest ← needy(SUpi), list(DPSvi)

Outputs measures:

1. Nps←No. Of people saved
2. Spv←Schedule of person to vehicle

The major objective here is to provide healthcare to the injured people due to accidents. The major factor is to determine the count of how many peoples are injured in the accident, out of the total count of the occupancy in the vehicle, in order to predict the number of people need healthcare assistance along with the severity of the accident. The above PAB-PSO algorithm consider the ratio of total life saved vs. total count need healthcare assistance. At the primary stage the algorithm rearranges the injured list in the sorted order to make priority list and based on the priority list the people will get assisted. Once the number of vehicles that are meet in an accident is predicted then the total count of people occupied in the vehicle need to predicted with the score ranges from 0 to 1 where 1 as full, between 0.5 partial and 0 as no involvement. After the local maxima (i.e., 1) reached the vehicle must reached to the hospital immediately. Whereas for the other score values, it continues assisting the peoples. At last, the algorithm provides the rate of people saved in an accident based on severity rate, total count of injured people and total count of people involved in an accident.

5. Evaluation of PAB-PSO (Person Accident based PSO algorithm)

In order to better understand the above describe and proposed PAB-PSO algorithm, we have considered the sample scenario: The 10 tasks are available to be scheduled on 10 already working processors with some tasks in a process as follows:

5.1 Task Scheduling Problem

Sr.No	Parameters	Values
1	Task	T01 to T10
2	Rate of emergency	0.5,0.3, 0.6,0.8, 0.7,0.9, 0.6, 0.3,0.5,0.4
3	Available processors	P01 to P10
4	Involvement rate / processor	0.2, 0.5, 1,0.5,0.5,1,1,0.6, 0, 0,
5	Probability of Processor Occupied	0.4,0.5, 0.6, 0.5,0.4, 0.6, 0.8, 0.2, 0.9, 0.2
6	Task capacity / processor	2, 3, 3, 2,4,3, 4, 1, 2, 3
7	Participation of each processor	0 / 0.5 / 1.0 [NO/PARTIAL/FULL]

Table 1. The number of officially reported plague cases in the world.

5.2 Algorithm contrivance coup

1. The list of tasks set to be scheduled as SUpi with the available processors need to be sorted in descending order of task priority as T6,T4,T8,T5,T3,T1,T9,T10,T2,T7
2. % processor empty: Evi = 0.6, 0.5, 0.4, 0.5, 0.6, 0.4, 0.2, 0.8, 0.1, 0.8
3. Tasks schedule capacity: ESvi = 1.2, 1.5, 1.2, 1.0, 2.4, 1.2, 0.8, 0.8, 0.2, 2 .4
4. Sorted ESvi = DESvi

   4a. Evaluating DESvi * Pvi = PSvi

   and sort in descending order to find DPSvi.

4b. Find the convergence factor (Cf) = (No. of people Assisted / Total number of people).

P10←2.4 :: P5←2.4 :: P2←1.5 :: P1←1.2 :: P3←1.2 ::

P6←1.2:: P4←1.0 :: P7←0.8 :: P8←0.8 :: P9←0.2

5. Pvi =1 for P2, P1, P3, P6, P4:

P2←1.5 :: P1←1.2 :: P6←1.2 :: P3←1.2 :: P4←1.0
T6 → P2:: T4 →P1 :: T8 →P3 :: T5→P6 :: T3→P4

If (urgency (Pi) > urgency (Pj) && if Pi volunteers for Pj) → {(urgency (Pi ⇔ Pj)) and Restart the complete system again and serve Pj before Pi.} (i.e, if particle Pi belongs to young and particle Pj belongs to child or elder then the system must serve Pj before Pi based on the criticality of the injury and reset the system)

6. DESvi * Pvi

P10←2.4:: P5←2.4 :: P7←0.8 :: P8←0.8 :: P9←0.2

T1→ P10:: T9→ P10:: T10→P5 :: T2 → P5 :: T7→ P8

This implies that task 1, 9 is scheduled on P10 processor and Task 10,2 on P5 and Task T7 on P8 processor.

6. Results

Figure 2, Shows the results of above described PAB-PSO algorithmic results, by simulating multiple times along with parameters: no. of peoples, no. of vehicles, vehicle capacity, and accuracy. Through the experimental analysis it has been observed that the highest accuracy seems to be 93% as we increase number of persons to serve.

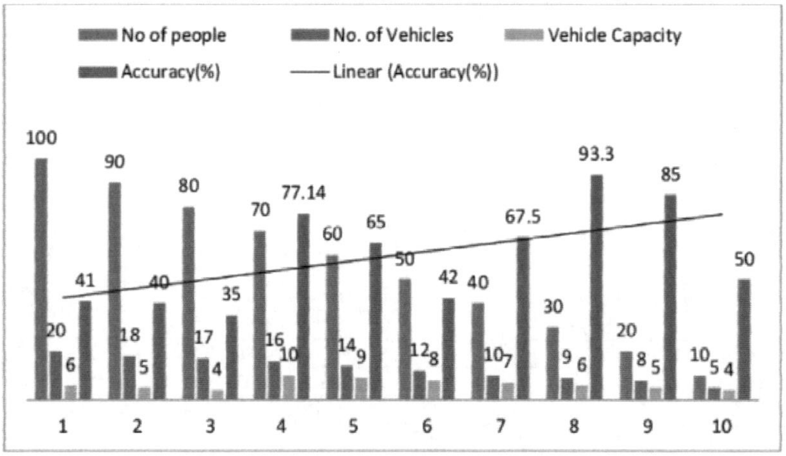

Figure 2. Runs of algorithm with change in values with effect on accuracy and delays.

Fig. 3a and 3b, Illustrates the performance analysis of the above described PAB-PSO algorithm. We have simulated the designed algorithm around 50 times and the observed the results of all benchmark functions CEC 2006 and presented all the attainable benchmark functions with their output characteristics found in the analysis of outcome from the experimental simulation. We have seen different characteristics of benchmark function as quadratic, non-linear and linear in nature.

IOP Conf. Series: Materials Science and Engineering 1049 (2021) 012010 doi:10.1088/1757-899X/1049/1/012010

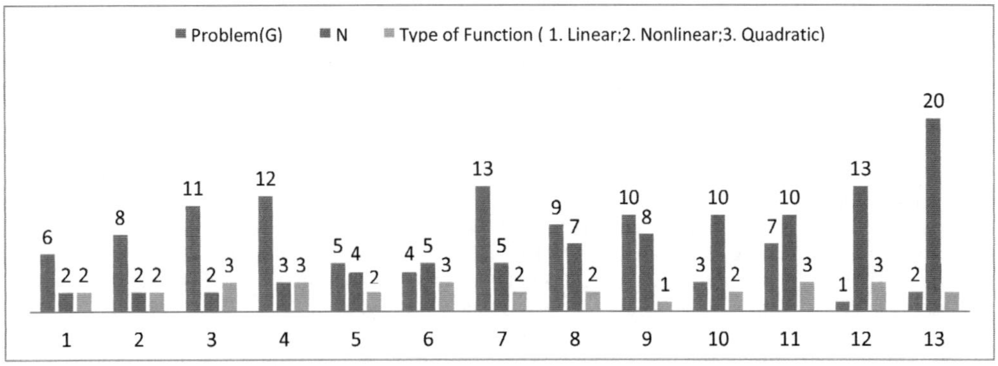

Figure 3a. Benchmark functions G01 to G013 characteristics

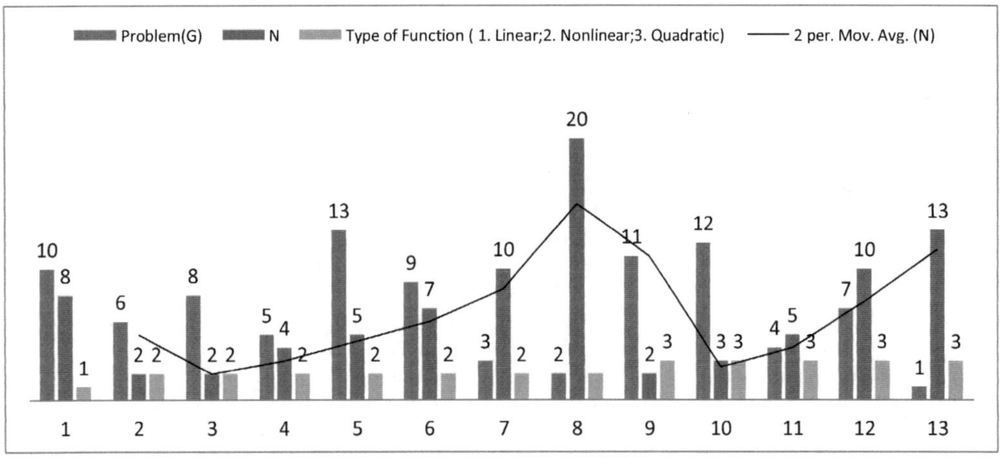

Figure 3b. Benchmark functions G01 to G013 characteristics

Table.1, Shows finding of experimental results based on multiple run (50 times) performed independently on each test functions G01 – G13 from the standard test functions of CEC-2006 test suit on "Matlab" as a simulating tool; Machine with T8100, 2.10 GHz, 3M cache processor and 2GB memory. The algorithmic simulation results on test function based on optimized value, accuracy, delay, no. of run, lowest generation is presented this table(1). The values where the global optimum is reached is represented with * mark and '-' represents where no results meet the desired criteria to be as candidate value. Out of G01-G-13 benchmark function: G1, G2, G3, G4, G8, G10, G11, G12, G13 reached to the level of being as global optimum.

Sr. No	$f(x)$	Optimization Function	Global Optimum solution $f(x*)$	Optimized value	Max Efficiency Accuracy %	Delay	Run No in which optimum found	Lowest generation
1	Min	G01	-15.000*	-15	100	1.197	2	-9
2	Max	G02	0.803*	0.729	34.74	0.019	9	0.191
3	Max	G03	1.000*	0.992	99.26	0.011	16	0.038
4	Min	G04	-30665.539*	-30679.001	99.95	0.019	9	-26897
5	Min	G05	5126.498	4358.770	85.02	0.022	16	252.608
6	Min	G06	-6961.814	-6440.203	96.54	0.012	5	-1575.297
7	Min	G07	24.306	159.843	15.20	0.026	16	159.843
8	Max	G08	0.095*	0.098	96.84	0.013	18	0.0003
9	Min	G09	680.63	988.283	68.86	0.022	26	-
10	Min	G10	7049.25*	7076.447	99.61	0.017	17	5361.759
11	Min	G11	0.75*	0.75	99.90	0.011	14	0.0122
12	Max	G12	1.000*	0.980	98.06	0.398	13	0.797
13	Min	G13	0.053*	0.053	99.94	0.491	11	0.044

Table 2. Obtained results from PSO algorithm

7. Conclusion

In this paper, we have presented an improved version of Particle swarm optimization named PAB-PSO to optimize the process of serving the healthcare facilities during the road accidents at the roadside. In the process of serving the people who met with road accidents, it is most important to provide life care assistance service immediately. In order to make this process optimized and ability to serve more and needier peoples at road side we have considered previous knowledge of person's accident behaviour as a learning model of swarm particles and make the process optimized to get maximum throughput. The proposed algorithm has been evaluated on G0-G13 benchmark function using MATLAB featured tools and the results have been presented. Out of the evaluated benchmark functions G01-G-4; G8; G10-G13 is reached to a global maxima. However the functions G5-G7 and G9 show worst performance during evaluation. We have also presented the various characteristic features of these functions shown as quadratic, non-linear and linear during evolution. The sample evaluation of the proposed algorithm is also discussed in section(6) which will helps us to understand how the proposed algorithm will going to schedule a processes and optimizes the complete healthcare assistance at road side . At the end it is important to compare these results with the available benchmark techniques for analysis and validation.

References

[1] Parsopoulos, K.E.Vrahatis, M.N., 2001. "Modification of the particle swarm optimizer for locating all the global minima, Artificial Neural Networks and Genetic Algorithms", Computer Science series, Springer, Wien, pp. 324–327.
[2] Parsopoulos, K.E., Plagianakos, V.P. 2001. Magoulas, G.D., Vrahatis, M.N., "Improving particle swarm optimizer by function "stretching" ", Nonconvex Optimization and Applications, vol. 54, ch. 3, pp. 445– 457, Kluwer Academic Publishers, The Netherlands.
[3] Y. Shi and R. C. Eberhart. 2001. Fuzzy adaptive particle swarm optimization," in Proceedings of the Congress on Evolutionary Computation, pp.101–106.
[4] J. Kennedy and R. Mendes, 2002. "Population structure and particle swarm performance," in Proceedings of the IEEE Congress on Evolutionary Computation, pp.1671–1676.
[5] W.-J. Zhang and X.-F. Xie, 2003. "DEPSO: Hybrid particle swarm with differential evolution operator, "in Proceedings of the IEEE International Conference on System Security and Assurance, pp. 3816–3821.
[6] A. Ratnaweera, S. K. Halgamuge, and H. C. Watson, 2004. "Self-organizing hierarchical particle swarm optimizer with time varying acceleration coefficients," IEEE Transactions on Evolutionary Com- putation, vol.8, no.3, pp.240–255.
[7] J.J.Liangand P.N.Suganthan, 2005. "Dynamic multi-swarm particle swarm optimizer," in proceeding sof the IEEE Swarm Intelligence Symposium (SIS'05), pp.127–132.
[8] Bird S., Li X., 2007. "Using regression to improve local convergence. In: Proceedings of the 2007 IEEE Congress on Evolutionary Computation, Singapore, pp. 1555–1562.
[9] Yare, Y., and Venayagamoorthy, G. K., 2007. "Optimal Scheduling of Generator Maintenance Using Modified Discrete Particle Swarm Optimization," Proceedings of the Symposium on Bulk Power System Dynamics and Control - VII. Revitalizing Operational Reliability, iREP, Institute of Electrical and Elec- tronics Engineers (IEEE).
[10] Y.-P. Chen, W.-C. Peng, and M.-C. Jian, 2007 "Particle swarm optimization with recombination and dynamic linkage discovery," IEEE Transactions on Systems, Man, and Cybernetics, Part B: Cybernetics, vol.37, no.6, pp.1460–1470.
[11] Nema, S., Goulermas, J., Sparrow, G., Cook, P.,2008 "A hybrid particle swarm branch-and-bound (HPB) optimizer for mixed discrete nonlinear programming", IEEE Trans. on Systems, Man, and Cyber- netics, Part A, vol.38, no.6, pp.1411-1424.
[12] Yang, X.S., 2009. "Firefly Algorithms for Multimodal Optimization", In: Stochastic Algorithms: Foundations and Applications. SAGA 2009. Lecture Notes in Computer Science, vol 5792. Springer, Ber- lin, Heidelberg.
[13] Z.-H. Zhan, J. Zhang, Y. Li, and H. S.-H. Chung, 2009. "Adaptive particle swarm optimization," IEEE Transactions on Systems, Man, and Cybernetics, Part B: Cybernetics,

vol. 39, no. 6, pp. 1362–1381.

[14] Rymut, B., Kwolek, B., 2010. "GPU-supported object tracking using adaptive appearance models and Particle Swarm Optimization", Proceedings of the 2010 international conference on Computer vision and graphics: Part II, ICCVG'10, Springer-Verlag, Berlin, Heidelberg, 227–234.

[15] Yang, X. S., 2010. "A new meta heuristic bat-inspired algorithm," in Nicso 2010: Nature Inspired Cooperative Strategies, pp. 65–74, Springer, Berlin, Germany.

[16] Engelbrecht, A.P., 2010. "Heterogeneous particle swarm optimization", in: Swarm Intelligence, Springer, pp. 191–202.

[17] Zhang, J., Pan, T.-S., Pan, J.-S., 2011. "A parallel hybrid evolutionary particle filter for nonlinear state estimation," in Proc. Robot, Vis. Signal Process. First Int. Conf., Nov, pp. 308–312.

[18] Awwad, O., Al-Fuqaha, A., Ben Brahim, G., Khan, B., Rayes, A. 2013. "Distributed topology control in large scale hybrid RF/FSO networks: SIMT GPU-based particle swarm optimization approach," Inter- national Journal of Comm. Systems, vol. 26, no. 7, pp. 888–911.

[19] Chunlu Wang Chenye Qiu Xingquan Zuo Chuanyi Liu, 2013. "An Accident Severity Classification Model Based on Multi-Objective Particle Swarm Optimization", In: IEICE Transactions on Information and Systems Vol.E97-D No.11 pp.2863-2871.

[20] Chen, R.-B., Hsu, Y.-W., Hung, Y., Wang, W., 2014. "Discrete particle swarm optimization for con- structing uniform design on irregular regions," Computational Statistics & Data Analysis, vol. 72, pp. 282–297.

[21] Pan,T.S., Dao, T.K., Nguyen, T.T., Chu, S.C.,2015. "Hybrid Particle Swarm Optimization with Bat Algorithm". In: Genetic and Evolutionary Computing. Advances in Intelligent Systems and Computing, vol 329. Springer, Cham.

[22] Kvasov, D.E., Mukhametzhanov, M.S., 2016. "One-dimensional global search: nature-inspired vs. lipschitz methods." AIP Conf Proc 1738(1):400012.

[23] Wang, X.; Q., 2016. "The Study of K-Means Based on Hybrid SA-PSO Algorithm", In Proceedings of the 2016 9th International Sun, Symposium on Computational Intelligence and Design (ISCID), Hangzhou, China, 10–11 December, pp. 211–214.

[24] Phung, M.D., Quach, C.H., Dinh, T.H., Ha, Q., 2017. "Enhanced discrete particle swarm optimization path planning for UAV vision-based surface inspection", In Automation in Construction, Volume 81, 2017, Pages 25-33, ISSN 0926-5805.

[25] Xia, X., Gui, L., He, G., Xie, C., Wei, B., Xing, Y., Wu, R., Tang, Y., 2017. "A hybrid optimizer based on firefly algorithm and particle swarm optimization algorithm", In Journal of Computational Science, ISSN 1877-7503.

[26] Javidrad, F.; Nazari, M., 2017. "A new hybrid particle swarm and simulated annealing stochastic optimization method." Appl

Efficiency and Green House Emission Response of Different Vehicles Operating on Various Driving Cycles

Deepali Puri[1,1] and Leena G[1,2]

[1]*EEE department,Manav Rachna Institute of Research and Studies,Faridabad,India*

[1]deepalipuri17@gmail.com, [2]leenag.fet@mriu.edu.in

Abstract. In today's scenario, energy security and global warming are considered as vital parameters of concern. The use of large number of conventional vehicles on roads has caused serious environmental pollution and they consume considerable amount of energy resources .Thus solution lies in switching to cleaner and greener energy alternatives, such as electrical energy. In this paper, conventional, electric and hybrid electric vehicles are compared on the basis of efficiency and greenhouse gas emissions. For efficient comparison, three different cycles, namely, Manhattan, Urban Dynamometer Driving Schedule and Arterial, are chosen. The simulation and computer modelling is implemented using the Advanced vehicle simulator (ADVISOR) platform. When Electric vehicle is tested on three different types of driving cycles, it exhibits highest efficiency and no tail pipe emission as compared to other vehicles. Thus the use of Pure electric vehicle eliminates both the problem of depleting energy resources and environmental pollution.

Keywords— Electric vehicles, Conventional vehicles, Hybrid vehicles, ADVISOR, Driving cycle.

1.INTRODUCTION

Vehicle, the most important necessity of human life, is used to transport people or cargo. Automobile plays an incredible role in economic and social growth of any country. Conventional vehicles uses major portion of renewable energy resources like petroleum, diesel, gas etc., to drive their wheels but these resources are limited in nature. Moreover, it causes considerable pollution in the form of CO2, NO*x*, particles matter [1]. Resorting to electric energy is the best option to reduce CO2,NOx, Hydro Carbon(HC) emission, which is the prime culprit in global warning[2]. Over the past few decades, there is significant development in automobile industry [3]. Only Electric and Low emission hybrid vehicles can meet the outlined criterion of developing ultra-low emission automobiles as mentioned in the national regulations, which ensure a systematic and sustained development of low emission land vehicles.By 2035, almost 85% of all the vehicles introduced in the market will be electrical vehicles [4,5]. On broad terms, electrical vehicles are classified as Pure Electric vehicles (PEV) and Hybrid electric vehicles(HEV). The pure electric vehicle utilizes only one source of energy, electrical energy while hybrid electric vehicle uses the combination of two different power sources i.e, internal combustion engine and batteries. In this paper, a broad comparison between conventional, hybrid and pure electric vehicles are done on the basis of overall efficiency and Co_2,No_x,HC emission. The simulation and computer modelling is implemented by using Advanced vehicle simulator (ADVISOR) tool from the DOE's national Renewable energy laboratory [6].The performance of pure electric, conventional IC engine and hybrid electric vehicles are compared in terms of efficiency and vehicle emission[7,8]. These performance parameters are evaluated on different driving cycles [9,10]. This paper is organized as: Section II explains the vehicle dynamics and driving cycle; Section III introduces ADVISOR, a simulating tool. Simulation is illustrated in Section IV followed by conclusions and result in Section V.

2. VEHICLE DYNAMICS AND DRIVING CYCLE

2.1 Vehicle Dynamics

Electric vehicle is the type of vehicle that utilizes only electrical energy to propel its wheels or uses battery to store energy. The most commonly used, conventional vehicle uses only internal combustion for its working while hybrid electric vehicle utilizes both internal combustion engine and battery to drive its wheel. The basic difference between conventional and electric vehicle lies in its power train system. The combustion engine in traditional vehicles is fuelled by fossil fuel whereas electric energy is the main source in electric vehicle and it mainly includes battery system, control system and motor system.

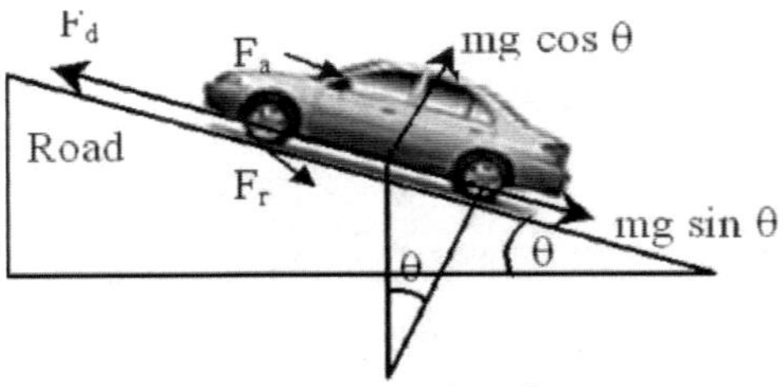

Figure. 1.Vehicle dynamics diagram

The vehicle modelling is based on the vehicle dynamic diagram as shown in figure 1 and is derived from the basic equation of Newton's Second Law, F=ma.
This equation can be modified with some explicit forces acting on vehicles

$$F = mgC_{rr} + {}^{1}/_{2}\,\rho C_D A V^2 + ma + mgsin\theta \quad (1)$$

Where, F= force required at the wheels of the vehicle, Newton
C_{rr}= coefficient of rolling resistance between tyres of vehicle and the surface of road.
m = mass of the vehicle, kg.
ρ = density of the ambient air, kg/m^3
C_D= coefficient of drag of the vehicle
A= area of cross section of the vehicle, m^2
v =velocity of the vehicle, m/s
a= acceleration of the vehicle, m/s^2
g= acceleration of gravity, m/s^2
θ= angle of inclination between vehicle and road, degrees
The first term specifies the essential force required for overcoming the rolling resistance of the vehicle. This force remains constant in spite of increase or decrease of the vehicle speed. The second term denotes the drag force that the vehicle has to conquer at a certain speed. The drag force is proportional to the square of speed of the driving vehicle. Third term specifies the inertial mass of the vehicle. It has no effect under the constant-speed conditions and is non-zero only when the vehicle is decelerated/ accelerated and the last term specifies the force required to propel the vehicle on a nonzero grade. Now the required torque, T(N-m) and angular velocity, w (radians per second)is calculated by using equations

$$T = Fr \quad (2)$$

$$w = \frac{v}{r} \quad (3)$$

Where v= linear velocity in m/s and r is the radius of wheel. The input angular velocity, $W_{in}$is calculated using the overall reduction ratio,$R_x$of the transmission. Compensating for the transmission efficiency, η_x as a function of torque-input speed, the input torque,T_{in} can be expressed as

$$W_{in} = W_{out} R_x \quad (4)$$

$$T_{in} = \frac{T_{out}\eta_x(T_{out}, W_{out})}{R_x} \quad (5)$$

The input power, P_{in} required from the vehicle voltage bus is calculated by using efficiency of inverter or motor at desired torque and speed

$$P_{in} = \frac{T_{out} W_{out}}{\eta_m(T_{out}, W_{out})\eta_i(T_{out}, W_{out})} \quad (6)$$

where $\eta_i$is efficiency of the inverter and η_m is efficiency of the motor .This power input is supplied either by batteries or auxiliary power units depending upon the type of vehicle used.

2.2 *Driving cycle*

The driving cycles are basically data point of speed (Km/hrs)versus time(seconds) that represent the particular driving sample. It is used to find out the performance of vehicle in terms of rate acceleration /deceleration, fuel consumption, emission, etc. A typical driving cycle curve is used for representing the progression of vehicle operating conditions; acceleration, steady state (const. speed) and deceleration for a given region, city or a country. To test the vehicle performance and to check whether a design is appropriate for desired application or not, driving cycles are used. This results in reduction of expense on road test and save the fatigue and time of test engineers.

3.ADVANCED VEHICLE SIMULATOR (ADVISOR)

The modelling and simulation of vehicles are done in ADVISOR. It operates in MATLAB/Simulink environment as MATLAB is known for its modelling flexibility, visualization and optimization toolbox. In MATLAB files, vehicle data is provided and models are developed in Simulink. ADVISOR works in backward facing model, which requires no driver behaviour. It simply answers the question like how each component in vehicle configuration must perform in order to meet the desired trace and the force required to move the vehicle are directly calculated from the speed trace and then it is translated into torque. The driving pattern is set as input and force required to accelerate the vehicle is calculated and this force is converted into torque which is passed through the drive train until a tractive force is calculated. This calculation approach is carried backward ,element by element, until the amount of energy or fuel required to meet the desired trace is calculated.

4. SIMULATION

In this paper, three different driving cycles were chosen for comparison, namely Manhattan, Urban Dynamometer Driving Schedule (UDDS) and Arterial. The speed-time characteristics of above mentioned driving cycles are as shown in figure 2. The Manhattan driving cycle is a low speed and frequent stops cycle. The UDDS driving cycle represents the city driving condition for light duty cycle testing whereas the arterial driving cycle computes fuel economy of the heavy duty vehicle.

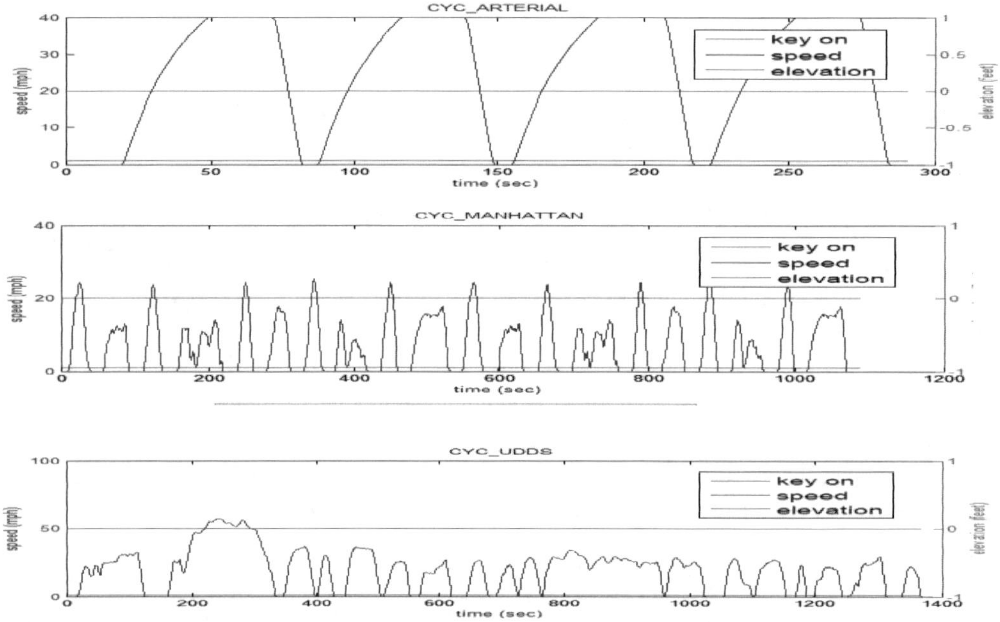

Figure. 2.Speed time characteristics of Manhattan, UDDS, Arterial driving cycles

The Electrical, hybrid and conventional vehicle are simulated in ADVISOR platform. Table 1 represents various parameters associated with vehicles and its specifications

Table 1: Parameters and its specification for vehicles

PARAMETER	SPECIFICATIONS
Frontal Area	2.174 m^2
Mass of vehicle	1368 kg

Aerodynamic drag coefficient	0.327
Overall gear ratio	6
Coefficient of rolling resistance	0.008
Type of motor	Permanent magnet DC motor
Maximum torque	248 Nm
Battery type	NiMH
Pack voltage	380 V
Capacity of battery	104Ah
Initial SOC	1
ICE power	50kw

The data flow in electric, conventional and hybrid vehicles are represented in the form of block diagrams shown in figure 3. The input to each block diagram is from drive cycle. The drive cycle block contain all the data of drive train components and it sends the desired speed trace to the next block i.e vehicle block. In each component model data is defined by the text files in the database. For example motor/control block contains all motor related data. The vehicle block has no drive train performance restrictions. It directly uses the desired trace for finding the average speed and force required over the time step. Now this information is passed to wheel/axle block which convert this linear speed and force into angular speed and torque using vehicle parameters. Depending upon the efficiency of the motor at any particular torque and angular speed, the input power required from the vehicle's voltage bus is computed. This required power is supplied either by the batteries or by combination of both auxiliary power unit and batteries. Once this calculation work has been done, software ADVISOR updates fuel used state of charge(SOC) of battery, emission etc., and then moves to the next step.

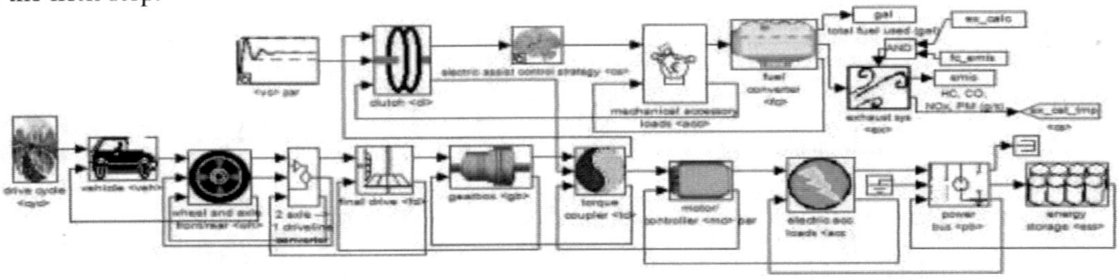

Figure 3.Block diagram of the hybrid electric vehicle using ADVISOR

The vehicle emissions computed using ADVISOR for the three different types of vehicles are tabulated in Table 2. Three categories of vehicle emissions i.e Carbon Monoxide, Oxides of nitrogen (NOx) and hydrocarbons are studied and shown in tabular form. The graphical representation of these emissions for hybrid, electric and conventional vehicles are shown in figure 4

Table 2: Green gas emission for electric, conventional and hybrid vehicles

Greenhouse gas emissions(grams/mile)	**Conventional Vehicle**	**Electric Vehicle**	**Hybrid Electric vehicle**
Oxides of Nitrogen (NOx)	1.89	0	1.606
Hydrocarbons (HC)	7.4	0	6.57
Carbon Monoxide (CO)	1.29	0	0.909
Total emissions(grams/mile)	**10.58**	**0**	**9.085**

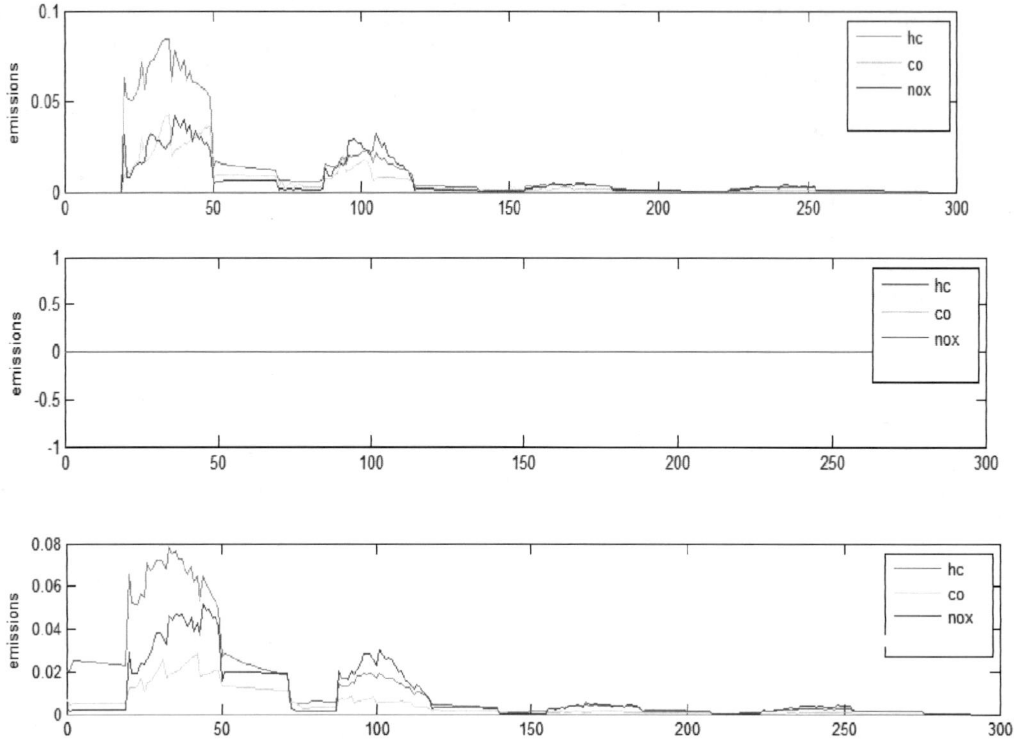

Figure 4: Emission graph for hybrid ,electric and conventional vehicles

Table 3: Calculation of overall efficiency for different driving cycles

Type of vehicle		Fuel Converter	Cluth	Torque Coupling	Energy Storage	Motor Controller	Gear Box	Final Drive	Wheel/ axle	Auxiliary Load	Aero	Rolling	Efficieny = Aero +rolling /fuel in-storage
Driving cycle-MANHATTAN													
Conventional Vehicle	In	12818	1549				1308	1225	1225	762			
	out	1900	1308				1225	1225	1132	0			
	Loss	10918	241				83	0	93	762	66	289	
	Efficiency	0.15	0.84				0.94	1	0.92	0			**0.027**
Electric Vehicle	In				162	2159	1491	1408	1408	762			
	out				2842	1491	1408	1408	1309	0			
	Loss				80	668	83	0	99	762	66	336	
	Efficiency				0.86	0.69	0.94	1	0.93	0			**0.145**
Hybrid Vehicle	In	11122	1574	1755	209	1055	1755	1645	1645	762			
	out	1574	1392	1755	1739	373	1645	1645	1537	0			
	Loss	9548	182	0	39	682	110	0	108	762	66	396	
	Efficiency	0.14	0.88	1	0.88	0.35	0.94	1	0.93	0			**0.036**

Driving cycle-ARTERIAL

Conventional Vehicle	In	6493	1391				1367	127 1	1271	204			
	out	1537	1367				1271	127 1	1180	0			
	Loss	4956	24				96	0	91	204	32 7	280	
	Efficiency	0.24	0.98				0.93	1	0.93	0			**0.09**
Electric Vehicle	In				299	1867	1541	142 2	1422	204			
	out				204 6	1541	1422	142 2	1323	0			
	Loss				120	326	119	0	99	204	32 7	325	
	Efficiency				0.8	0.83	0.92	1	0.93	0			**0.34**
Hybrid Vehicle	In	6104	1224	17 27	360	841	1727	161 5	1615	204			
	out	1224	1208	17 27	102 2	519	1615	161 5	1507	0			
	Loss	4880	16	0	74	322	112	0	108	204	32 7	384	
	Efficiency	0.2	0.98	1	0.83	0.62	0.93	1	0.93	0			**0.1**

Driving cycle-UDSS

Conventional Vehicle	In	22570	4219				4118	382 2	3822	958			
	out	4822	4118				3822	382 2	3534	0			
	Loss	17748	101				296	0	288	958	10 56	1042	
	Efficiency	0.2	0.98				0.93	1	0.92	0			**0.093**
Electric Vehicle	In				440	5947	4643	427 6	4276	958			
	out				678 1	4643	4276	427 6	3964	0			
	Loss				279	1304	368	0	311	958	10 56	1211	
	Efficiency				0.85	0.78	0.92	1	0.93	0			**0.34**
Hybrid Vehicle	In	22212	4888	52 03	580	1610	5203	486 2	4862	958			
	out	4888	4799	52 03	243 9	604	4862	486 2	4521	0			
	Loss	17323	90	0	96	1006	341	0	341	958	10 56	1429	
	Efficiency	0.22	0.98	1	0.87	0.38	0.93	1	0.93	0			**0.103**

Three different kinds of vehicles, conventional, hybrid and electric vehicles are run on different driving cycles namely, Manhattan, urban dynamometer driving schedule and arterial. Step by step, energy in, energy out, losses and efficiency of each component is calculated and tabulated as shown in table 3.At the end, overall efficiency of all the types of vehicles considered in this paper is calculated for the different driving cycles.

5. RESULT AND CONCLUSION

The simulation results reveals that when electric , hybrid and conventional vehicles are compared for different driving cycles namely, manhattan, UDDS and arterial , the efficiency of electric vehicle for each of the above driving cycle is more as compared to other types of vehicles. The average efficiency of conventional vehicles is 0,070 , hybrid vehicle is 0.079 whereas for electric vehicle it is much higher that is 0.275.When these vehicles are compared on basis of greenhouse emission it is found that the conventional vehicle emits 10.58 grams/mile ,hybrid vehicle emits 9.085 grams/mile and electric vehicle emits zero grams/mile. So, economic and social status of any country can be raised by switching over to greener and cleaner solution i.e Electric vehicle.

6.References

[1] Bolaji B and Adejuyigbe S 2006.Vehicle emissions and their effects on natural environment – a review .J ghana inst eng. **4(1),** 35–41

[2] Hawkins T ,Gausen, M and Strømman,A,2012. Environmental impacts of hybrid and electric vehicles—a review . *Int j Life Cycle Ass*.**17(8)**, 997–1014.–14

[3] Jain R 2018. The progress of electric vehicle. *Int Res J Eng Technol*.**5(10)** , 1613-19

[4] Global Electric Vehicle (EV) Outlook 2018 IEA Technical report, May 2018 , Paris https://www.iea.org/reports/global-ev-outlook-2018

[5] Rajashekara K 2003.Present Status and Future Trends in Electric Vehicle Propulsion *.IEEE.* **1(1)**, 3-10

[6] Markel, T Brooker, A Hendricks, T Johnson, V Kelly, K Kramer, B O'Keefe and N M 2002 ADVISOR: a systems analysis tool for advanced vehicle modeling. *J.Power Sources* .**110(2),** 255-66.

[7] Sam A Grossman D and ParkJ 2010.A study on optimization of hybrid drive train using advanced vehicle simulator. *J.Power Sources*.**195(19)** 6954-63

[8] Hassan,A Mushtaq Z and Rehman, A2015 Performance Analysis of Drive Train Configurations using ADVISOR. *Int. j. innov. res. adv. eng* . **2(2)** 306-12

[9] Chew K Koay C Yong, Y 2014 ADVISOR simulation of Electric Vehicle Performance on Various Driving Cycles. *Int. j. innov. sci. eng. technol*.**1(8)** 70-76.

[10] Shah V Pritesh P Sagar Pand Maheshwari R 2011 Measurement of Real Time Drive Cycle for Indian Roads and Estimation of Component sizing for HEV using LABVIEW. *In J of Elect Comp Eng* **5(10)** 1112-21

[11] DieselNet,n.d.EmissionTestCycles.[online]Dieselnet.com. http://www.dieselnet.com/standards/cycles/

[12] Barlow T Latham S McCrae I and Boulter P 2009 A reference book of driving cycles for use in the measurement of road vehicle emissions, London:TRL, ISSN: 0968-4093

[13] Yong J Ramachandaramurthy K and Mithulananthan N 2015 A review on the state-of-the-art technologies of electric vehicle, its impacts and prospects *Renew. Sustain. Energy Rev* **49**, 365–385

Sentiment analysis on micro-blog data using machine learning techniques - A Review

M S Kalaivani[1], S Jayalakshmi[2]

Tagore College of Arts and Science, Chennai,600049,TamilNadu,India
VELS University,Chennai, 600117, Tamilnadu,India

Email: kalaims2007@gmail.com,jai.scs@velsuniv.ac.in

Abstract :The user generated content in social networks, websites, ratings and reviews about product, people , places have become very beneficial for various fields to improve its growth. Sentiment analysis, which studies user generated content and find the correct sentiment polarity. Data collection, feature extraction, sentiment classification and decision making are the steps involved in this process. Machine Learning and Deep learning Methods are used to predict the sentiments. Large Input dataset from websites or microblog is taken as input and processed. There are some challenges in identifying accurate decisions. This Review Paper focus on recent studies about learning models such as Support vector Machines ,Maximum Entropy , deep learning models and various methods which are used in solving sentiment classification, product review analysis, Text Processing and visual analysis etc.,

Keywords : sentiment analysis, Machine Learning ,Deep Learning models, Support vector Machine, Convolutional Neural networks.

1. Introduction

Nowadays, enormous amount of information is presented in webpages, databases, and blogs as online document.The unstructured form of data from the internet required to be changed well-structured data, toanalyze further. The Analysis process used in many fields such as Consumer information, websites, and Social Media [1].The goal of sentiment analysis is study user feedback and predict the sentiments. Usually, It defines opinion polarity, i.e., if somebody has positive, negative, or neutral opinion towards something. For example, (i) if a person plan for a trip to a particular place, he simply refer the online feedback of previous visitors for making the conclusion.

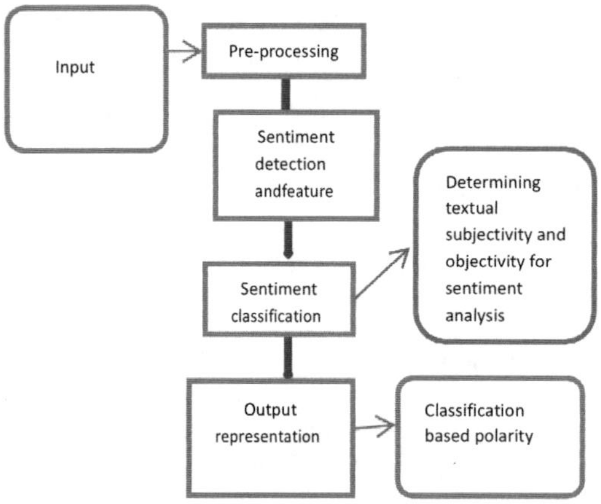

Figure 1. Process of sentiment Analysis

(ii) If a person want to purchase a product, he reads online ratings and comments first, and take a decision to buy the product or not. The need of sentiment analysis is raised because requirement of analyzing and formatting the unstructured data in social media or websites .[2].The idea of sentimentanalysis is , without human intervention it determines the way of user reviews [3].

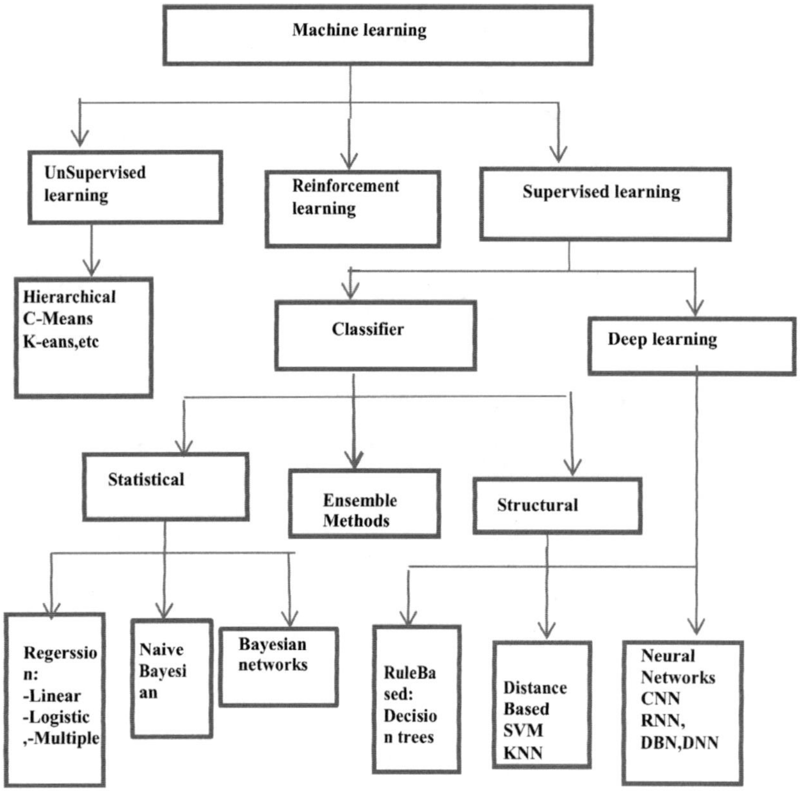

Figure 2 .Types of Machine learning

In social media, Twitter has made a big revolution, which is connecting millions of people worldwide through micro blogging service. The registered user of twitter can post short messages about anything, these short messages are called Tweets, which can have character-set length upto 280 .In the recent years researchers applied SA techniques with twitter comments and the results helpful in decision making.Machine learning techniques are applied by extracting the sentences and features. The polarity score has been assigned to the lexicons and sum of the polarity score is calculated from the overall values. [4]

1.1.Sentiment analysis with product review

The human's opinions, sentiments towards a product are being analyzed by the machine and the results encourage the purchasing decision of the customers .product review analysis is an entirely recognized field, where the customers can easily decide their product needs. Initially the product reviews will be taken as input text, and preprocessing is done .In this phase the input file is normalized, which ready for training process. This file is mapped with the saved features for training. After mapping the data will be fed into the different models and the performance of the models are analyzed as results. Several Machine Learning and are proposed in the past .Deep learning models, a division of Machine learning well suited for sentiment analysis with large dataset. The following is the basic procedure in sentiment analysis for product review.

Procedure for SA process

Step 1: Input text paragraph from dataset
Step 2: Preprocessing the text. It includes removing stop words, removing numbers , expanding acronyms, replacing negation and removing URLs .
Step3 :Feature Extraction phase: Extracting words corresponds to adjective, adverb and verb.
Step 4:Selection of essential features for training the input data. Mapping the input data with selected features.
Step 5:The dataset can be divided 70% training and 30% testing.
Step 6: Training dataset to Machine learning algorithms or rain the model with Deep learning techniques
Step 7: Test the model with different values.
Step 8: Result analysis with performance of the model.

2. Literature Review

In this section various studies about sentiment analysis by using MachineLearning and Deep Learning models are explained. Naive Bayes, Maximum Entropy, Ensemble methods, SVM,Decision trees, Stochastic Gradient descent are some of the familiar machine learning methods used in recent researches.

.2.1. Machine Learning Techniques

The Author [5,MonishaKanakaraj], analyzed Navies Bayes, MaximumEntropy, Support vector Machines algorithms in his work. Twitter dataset 7086 reviews from election dataset, 25000 reviews from movie dataset were taken as input for sentiment analysis. Randomized Trees classification method produces better outcomes . Author [6, Ana Valdivia], applied sentiment analysis with Trip advisor; it is an American travel website company that provides user opinions about their hotels, restaurant and monuments. The analysis task categorized into three steps, polarity detection, Aspect selection, and classification. The author applied SAMs , SentiStrength, Bing,Syuzhet, and CoreNLP methods to analyze the user opinios.The result shows that Syuzhet detect at best 57.48 and 45.10 percent of negative feedback.In this paper [7, (Geetika Gautam)] Author proposed machine learning techniques for sentence classification and taken product reviews as input dataset from twitter data.
The naïve byes method provides better result than maximum entrophy.SVM provides improved results in unigram model. The accuracy level is gained in semantic analysis word net to to 89.9% from 88.2%.

In this study [8, (Neethu M S)] Author, Proposed sentiment analysis on electronic products like mobile, laptops by using machine learning algorithms. They presented a new feature vector for classifying the reviews into positive and negative. User reviews about electronic products taken from twitter as input dataset and performed sentence level sentiment analysis.Navie Bayes, SVM, Maximum Entropy Classifier and Ensemble classifiers are applied and performance analysed.The result shows NaiveBayes has 89.5 % accuracy whereas remaining three methods obtained 90% accuracy level.The Author [9, (Venkata Sasank Pagolu)] predicts the stock market movement of a company by using supervised machine learning algorithms. The classifier Word 2vec representations of tweets trained on Random Forest algorithm. The result shows 70.2% of accuracy with Random forest algorithms and with N-gram classifier it shows the accuracy level up to 70.5%.

In this study[10,Sankar] has proposed sentiment analysis about movie review by taking inputs from thee datasets Polarity, IMDb, and Rotten Tomato and applied the CNN approach in a mobile environment. The result shows that GloVe word embedding performs better. The GloVe 6B and Glove Crawl carry out in the same way. The SSWE word vector not produced expected outcome. In classification part, CNN and trained word vectors works better in mobile applications.

Author[11,(Akshi Kumar)] specified approach for sentiment analysis on twitter data. User opinions extracted withadjectives, verbs, adverbs from tweets. To get the semantic orientation of adjectives, corpus-based approach was applied It has been considered a motivational technique for upcoming researches.In this work, Author [12, (Akshat Bakliwal) used NLP techniques NGram ,POS-Tagged for sentiment analysis .Movie reviews and product reviews are taken as input dataset. Multi-Domain Sentiment Dataset (Version 2.0) has product reviews taken from Amazon.com with many (total 25) categories. They considered 4000 positive and 4000 negative reviews. Movie Review Dataset has 1000 positive and 1000 negative reviews. Machine learning algorithms SVM, MLP and NB applied and results were analysed.In the N-gram technique SVM achieved better results in both datasets 81.15% and 79.4%n POS-tagged approach MLP achieved 81.60% and 79.47% for movie reviews and product reviews.

In this study [13, (Chris Jefferson)] proposed fuzzy rule based system for sentiment analysis. Fuzzy membership degress provides more refined outputs. Decision trees and Naivebayes methods compared with fuzzy rule based system. All classifiers achieved good results with accuracy level 0.9. The fuzzy rule based system obtained better performance than other classifier.In this study the author[14, (A. A. Aliane1)]proposed a genetic algorithm for Arabic sentiment analysis by taking the input dataset from book reviews. They have performed five basic algorithms SVM, Multinomial Naive Bayes (MNB), Naive Bayes (NB), Stochastic Gradient descent, DecisionTrees and the outcomes were compared. The research shows that proposed algorithm performs better than previous classifiers. MNB obtained 94% accuracy level with unigrams .

In this study author [15, (Zhao Jianqiang1)] evaluated various pre-processing methods on five different input dataset.To identify the sentiment polarity two feature model- Word n-grams features model and Prior polarity score feature model were used. The classifiers SVM, NB, LR, RF, and baseline method (C-Method) were applied. The result shows that, NB and RF classifiers are more efficient than LR and SVM classifiers.

IOP Conf. Series: Materials Science and Engineering 1049 (2021) 012012 doi:10.1088/1757-899X/1049/1/012012

Table 1 Analysis of Machine Learning Techniques in sentiment analysis.

Researcher Name and Year	Model Used	Objective	Data Set	Results	Challenges for Future work
MonishaKanakaraj and Ram Mohana Reddy Guddeti[5]	Ensemble methods and machine learning methods	Sentiment analysis on movie and election reviews	Twitter dataset 7086 reviews from election dataset,25000 reviews from movie dataset	Ensemble methods performed 3-5% better than other methods	Dataset used for training and testing can be in different ratio with large subsets,which is randomly selected.
Valdivia, M. Victoria Luzón, and Francisco Herrera[6]	SentiStrength, Bing,Syuzhet, and CoreNLP Methods	sentiment analysis with Trip advisor	Dataset taken from American travel website	Syuzhet detect at best 57.48 and 45.10 percent of negative reviews	The values are varying in user ratings and Evaluations of SAM .A new approach is required for efficient classification.
GeetikaGautam,Divakar yadav[7]	Machine learning techniques with semantic analysis	Sentiment analysis on Product review	Twitter dataset	The accuracy level is gained in semantic analysis word net to to 89.9% from 88.2%.	The training data can be enlarged for better results in sentence identification process.
Neethu M S,Rajasree R[8]	Navie Bayes, SVM, Maximum Entropy Classifier and Ensemble classifiers	sentiment analysis on electronic products like mobile, laptops	Twitter dataset	Naive Bayes has 89.5 % accuracy whereas remaining three methods obtained 90% accuracy level.	The input dataset can be can be increased ,All these classifiers havealmost similar accuracy for the new feature vector.
A. A. Aliane H. Aliane, M. Ziane N. Bensaoul[14]	SVM, Multinomial Naive Bayes (MNB), Naive Bayes (NB), Stochastic Gradient descent Decision Trees	a genetic algorithm proposed for feature selection in Arabic sentiment analysis	book reviews	Proposed algorithm performs better than previous classifiers. MNB obtained 94% accuracy level with unigrams.	Multi domain dataset can be taken as input dataset to apply feature selection based approaches.

IOP Conf. Series: Materials Science and Engineering 1049 (2021) 012012 doi:10.1088/1757-899X/1049/1/012012

2.2.Deep Learning Techniques

In deep learning, data representation is referred by increasing the abstraction levels . [16].In the previous papers, there are various deep learning Architectures are Discussed,. Deep learning is used effectively in both supervised and unsupervised algorithms.Author [17, Islam.J], proposed a CNN framework for predicting the sentiments from social images. This is implemented by using s using Caffe and Python on a Linux platform. This model can be used in multimedia content to know user feedback, predictive modeling and advertising. A twitter dataset with 1269 images is taken as input and back propagation method is applied.The proposed model achieved better performance than existing systems. AlexNet was used in earlier systems and GoogleNet provides nearly 9% improved performance.

Theauthors[18, A. Severyn]implemented deep learning methods with twitter data. To initialize the word embedding neural language is used.Embedded words and parameters are applied to initialize the network, and training phase done by using supervised corpus Semeval-2015. Sentence matrix pooling, activations, convolutional layers and softmaxare the components used in proposed system. Deep learning technique is implemented in messagelevel, phrase level to predict polarity and better outcomes achieved.

The researcher [19, L. Yanmei] has explained about sentiment analysis with user opinions from a micro-blog. The user feedback on hot events analyses by using CNN . The corpus which has 1000 micro-blog comment divided into three **labels**. 274 for neutral , 300 for negative and 426 for positive emotions. The proposed work is compared with earlier models CRF, SVM, other traditional methods and performances analysed.The proposed model achieved the good performance in emotion analysis.

This study presents [20,ZeyuXiong] the CBOW model and CNNs to form a new deep learning model for paragraph vector representation. Stanford Sentiment *Treebank* dataset is used as input. Amazon Mechanical Turk used generates labels and 10000 paragraphs from Treebank taken as input. The training process needs the 56×50 matrix foreach paragraph .result shows that CBOW and CNN model achieved better results.

This study [21, Yongping Du] targets at sentiment classification of short messages. The author proposed capsule-based hybrid neural network model to attain the semantic data efficiently. In this model, BGRU (Bidirectional gated recurrent unit) applied to handle interdependent features with long distance. It extracts text to improve quality of the expression. There are different input dataset used in this system.
(1) Movie Review Data (2) NLPCC 2014 Data, 10000 product reviews. The performance of the system is compared with attention-based model and CNN (convolutional neural networks) model. The result shows that, movie review data achieved the highest accuracy of 0.8255 than NLPCC2014.

In this study [22, R. Socher], RNTN (Recursive Neural Tensor Network) and Sentiment Treebank model are projected to explain the compositional effects at various sentence levels. In previous models, the meaning of long phrases could not be analyzed well by the tool semantic word spaces. Effective training resources are needed for sentiment detection. The RNTN achieved accuracy level of 80.7% ac in sentiment predictionThis author [23, T. Chen] developed a sequence model for embedding of reviews of products which has temporal nature. Sentiment analysis is done with machine learning classifiers by taking three datasets as input from Yelp and IMDB.Every review is labeled with respect to rating score. Back-propagation algorithm with Adam stochasticoptimization methods are applied in training phase. The result shows that the proposed model performs better in document level sentiment classification .

Table 2 Analysisof Deep Learning Techniques in sentiment analysis.

Researcher Name	Model Used	Objective	Data Set	Results	Challenges for future work
ZeyuXiong[20]	CBOW model and CNN	paragraph vector representati on	10000 paragraph from Stanford Sentiment *Treebank* dataset	CBOW and CNN model achieved better results.	Skip-gram and CNNs can be combined to produce paragraph vector.
YongpingDu,Xi aoz hengZhao, Meng He - Wengyanguo[21]	capsule-based hybrid model	sentiment analysis social of data	Movie Review Data, NLPCC 10000 product reviews .	Movie review dataset obtained the highest accuracy of 0.8255.	This model can be extend to aspect-based opinion classification
Socher R, Pere- yginA, and Wu J, [22]	RNTN	Semantic. Composition ality	11,855 sentences from movie review (Pang and Lee 2005)	The RNTN provides 80.7% accuracy with,9.7 % over baselines	The training and testing datasets can be increased and various classification methods can be added .
Chen T,Xu, R. He. Xia, and Wang , [23]	Recurrent Neural Network(RN N- GRU)	User and Product Distributed Represent-Tions	Input dataset from Yelp and IMDB.	Proposed model outperform-ed than other models	sequence learning model, can be explored to bidirectional RNN and LSTM

3. Conclusion

The need of normalizing the online data increases researches in field of sentiment analysis. It extracts the unstructured data from social media and examines the user opinions. Machine Learning and Deep Learning Techniques play vital role in pre-processing the texts, classification, training, testing the models and result analysis. The outcomes of the research used to know the user expectations and feedback about a product, place, or any common thing. This analysis provides the way for development in all the fields. In this paper various studies have been discussed about machine Learning and Deep Learning Models which are used in sentiment classification.

4.Analysis with future work

This paper explained about some recent researches in sentiment analysis . There are so many models established by using Machine learning and Deep learning Methods. NaiveBayes, Support Vector Machines. Maximum Entropy, Ensemble methods, Decision trees, Stochastic Gradient descent are most used machine learning algorithms. Convolutional neural networks-CNN, DBN,Recursive neural network - RNN,DNN,RNN are frequently used Deep learning Algorothms. Since Deep Learning models have more hidden layers, performs better than other Techniques. These models provide efficient training in supervised and unsupervised ways and also it performs automatic feature extraction.

These models have some challenges, it requires large input dataset for better performance and it is costly to train. The performance of the algorithm varies depend on the problem and the input dataset.As a future work product review analysis can be done by using Deep learning Methods and the accuracy of the outcomes can be shown better than previous models.

References

[1] DoaaMoheyEl-Din Mohamed Hussein, 2016, *Journal of King Saud University - Engineering Sciences* survey on sentiment analysis challenges, Computers and Information Cairo University, Cairo, Egypt.
[2] M. Haenlein and A. M. Kaplan 2010 An empirical analysis of attitudinal and behavioral reactions toward the abandonment of unprofitable customer relationships,J.Relatsh.Mark.,vol.9,no.4,.
[3] F. Luo C. Li and Z. Cao, 2016 IEEE 20th Int. Conf. Comput. Support. Coop. Work Des.,Affective-feature-based sentiment analysis using SVM classifier, pp. 276281.
[4] Olga Kolchyna1 Th´arsis T. P. Souza1, Philip C. Treleaven12 and Tomaso18 Sep 2015 Aste12,arXiv:1507.00955v3 [cs.CL] Twitter Sentiment Analysis: Lexicon Method, Machine Learning Method and Their Combination
[5] MonishaKanakarajand Ram Mohana Reddy Guddeti2015 3rd International Conference on Signal Processing, Communication and Networking (ICSCN) NLP Based Sentiment Analysis on Twitter Data Using Ensemble Classifiers.
[6] Ana Valdivia M. Victoria Luzón, and Francisco Herrera University of Granada, IEEE INTELLIGENT SYSTEMS Sentiment Analysis in TripAdvisor, Published by the IEEE Computer Society.
[7] GeetikaGautam,Divakaryadav, 2012 IJCSI International Journal of Computer Science Issues, Sentiment Analysis of Twitter Data Using Machine Learning Approaches and Semantic Analysis, Vol. 9, Issue 4
[8] Neethu M S,Rajasree R, IEEE - 31661,4th ICCCNT 2013 July 4 - 6, 2013, Sentiment Analysis in Twitter using Machine Learning Techniques.
[9] VenkataSasankPagolu,KamalNayan Reddy Challa,GanapatiPanda,BabitaMajhi, 2016 Intelligent sentiment analysis approach using edge computing-based deep learning technique Sentiment Analysis of Twitter Data for Predicting Stock Market Movements.
[10] H. Sankar1 V. Subramaniyaswamy1V.Vijayakumar2 SangaiahArun Kumar3R. Logesh1 A. Umamakeswari1, *Softw: PractExper.* 2019;1–13. wileyonlinelibrary.com, John Wiley & Sons.
[11]Akshi Kumar and Teeja Mary Sebastian,2012 IJCSI International Journal of Computer Science Issues, Sentiment Analysis on Twitter Vol. 9, Issue 4, No 3
[12]AkshatBakliwal,AnkitPatil,PiyushArora,VasudevaVarma, 2011 Proceedings of the Workshop on Sentiment Analysis where AI meets Psychology (SAAIP), IJCNLP Towards Enhanced Opinion Classification using NLP Techniques, pages 101–107.
[13]ChrisJefferson, Han Liu IEEE and MihaelaCocea 2017 IEEE, Fuzzy Approach for Sentiment Analysis 978-1-5090-6034-4/17/ 2017 IEEE.
[14]Aliane H. Aliane M. Ziane1 N. Bensaou1 2016 IEEE A Genetic Algorithm Feature Selection Based Approach for Arabic Sentiment Classification, 978-1-5090-4320-0/16
[15]Zhao Jianqiang1 GuiXiaolin 2016 IEEE Comparison Research on Text Pre-processing Methods on Twitter Sentiment Analysis, 2169-3536
[16]Mufti Mahmud Mohammed Shamim Kaiser Amir Hussain and Stefano Vassanelli 2018 IEEE Applications of Deep Learning and Reinforcement Learning to Biological Data, 2162-237X
[17] Y. GuoY. Liu A Oerlemans S. Lao S. Wu and M. S. Lew, 2016 Neurocomputing, Deep learning for visual understanding: A review, vol. 187,pp. 2748

IOP Conf. Series: Materials Science and Engineering 1049 (2021) 012012 doi:10.1088/1757-899X/1049/1/012012

[18]J. Islam and Y. Zhang, 2016 IEEE International Conference on Big Data Cloud Comput. (BDCloud), Visual Sentiment Analysis for Social Images Using Transfer Learning Approach, pp. 124130.
[19]A Severyn A. Moschitti, 2015 Proc. 38th Int. ACM SIGIR Conf. Res. Dev.Inf.Retr.-SIGIR15,Twitter Sentiment Analysis with Deep Convolutional Neural Networks, pp.959962,2015.
[20]ZeyuXiong, QiangqiangShen, Yijie Wangl and ChenyangZhu,Paragraph 2018 CMC Vector Representation Based on Word to Vector and CNN Learning, vol.55, no.2, pp.213-227.
[21]YongpingDu,XiaozhengZhao,Meng He ,WengyangGuo, 2019 IEEE A Novel Capsule Based Hybrid Neural Network for Sentiment Classification , 2169-3536
[22]R. Socher A. Perelygin, J. Wu, 2013 Recursive deep models forsemantic compositionalityoverasentimenttreebank,Proc.,pp.16311642.
[23]T. Chen, R. Xu, Y. He, Y. Xia, X. Wang, 2016 Learning User and Product Distributed Representations Using a Sequence Model forSentimentAnalysis,no.August,pp.3444.

Medicine Distribution Robot and Human Less Intervention for Covid-19 Affected People (AKM MED ASSISTIVE BOT)

Kalpana Murugan [1], Md Arbas Ali Khan [2], MR Kylash [3], M Muralidharan [4]

[1] Department of Electronics and Communication Engineering, Kalasalingam Academy of Research and Education, Krishnan Koil, Virudhunagar-626126, Tamilnadu, India

[2,3,4] Department of Electronics and Communication Engineering, Kalasalingam Academy of Research and Education, Krishnan Koil, Virudhunagar-626126, Tamilnadu, India

E-mail: drmkalpanaece@gmail.com

Abstract. Social robots and administration robots with social insights are beginning to coordinate into our regular day to day activities. Managing health care needs of the patients collaborate with person-centered healthcare. Untimed medications to patients can cause serious health issues for the patients in hospitals and in old age nursing homes. The proposed work is intended to assist these patients with providing or recommending the necessary medication in the correct order at the specific times of the day without the actual need of a human nurse. This is incredibly tricky for the old patient who have issues in monitoring their medication at regular intervals. Thus, to avoid this, we have developed this medication robot. Which is a semi- autonomous microcontroller based programmable robot that works on the 'line following robot' concept, which can remind the patients, to consume their pills at regular intervals in a day. After successful intake of pills, the status of the medication is consequently sent to the patient's relative through an SMS.

1. Introduction

Robots are a needed and fast growing part of the modern medical society. The Robots has their history in medical field from the period of 1985.They are becoming unavoidable in this modern world. Propels in sensor and movement control advances cruel robots are more exact and independent than ever, and competent of not fair helping, but carrying out complex assignments themselves. The drugs that have been endorsed by medical specialists and doctors are bound to be successful on the off chance that you adhere to the specialist's definite directions on when and how to take them. The health specialist needs to endorse a specific medicine since the person feels that it will improve the patient's condition. This medicine is bound to be viable on the off chance that you follow your prescription course as endorsed. Inability to do so could, in certain conditions, have perilous consequences [1].

The step by step exercises of clinical laboratories will be fundamentally influenced by two problematic advances: computerization and the man- made thinking (the PC systems prepared to perform tasks that normally require human knowledge). These progressions will similarly broaden the degree of research in medical field. Computerization will realize extended adequacy yet will anticipate that changes should explore office establishment and a move in workforce getting ready requirements or mechanization in nursing care [2]. On the off chance that HNRs (Human Nutrition Research) are to help patients legitimately,so there should be a similar degree of extensive results capacity and the ability to respond as that of human medical caretakers. This incorporates capacities to truly watch, judge, quickly react, and lead human caring work on accentuating individuality [3]. The innovation utilizes a plan to assist patients with monitoring their medication utilization through a progression of light-transmitting diode (LED) caution marker signs and sound alert pointer signals. In the event that the patient isn't following the treatment that was recommended to the person in question, the framework can send tolerant consistence data via phone to an illness monitory board system [4]. The quantity of meds to be taken by every individual has expanded. It has gotten difficult for us to remind ourselves to take the medications at specific time. The new anticipated element in our task is our framework is reasonable that patient has taken medication or not and in this way the patient can't delay the time on which he needs to take pills.

1.1 Challenges

While implementing this project, there were some challenges like Learning Curve, The Cost of new technologies, Security, Simpler User Interface, Employment Issues, Poor visual perception.

1.1.1 Learning Curve: Practicing a new medical technology on a patient has more risk factors included in it. Medical experts have been told to put off utilizing the modern and progressed gadgets without, to begin with considering the mindfulness, preparation and instruction, required to create beyond any doubt the security of patients.

1.1.2 The Cost of new technologies: The Advanced new innovations in medical field can make the cost of the medical equipment's to increase every year, but can also increase efficiency.

1.1.3 Security: The most sought after and main thing for every healthcare application is the protection of trusted health data of any patient. The healthcare industry is still in a look for progressed security highlights.

1.1.4 Simpler User Interface: Generally, a less difficult and client inviting interface increments the chances of an application in getting embraced more within the environment utilizations. Therefore, a user-friendly frameworks could be a need in healthcare Types of gear and Gadgets.

1.1.5 .*Employment Issues*: Basically, robots are defined with a greater proficiency and production at a cheap cost. It may make a way for the lay off in all the sectors for the existing of Robotics and it becomes the major crisis.

1.1.6 Poor visualperception: This is one of the major factor for medication utilization blunders or medical errors due to forgot to take the medicines, for example, mal-dose, since the older people problematic to peruse the guidance on the medication case. The above challenges are overcome by our project.(i.e) AKM Med Assistive Bot.

2. Literature Survey

Human-Robot Interaction (HRI) is a field identifying with getting, arranging, and surveying mechanized systems for use by or with individuals. Correspondences between a human and a robot may take on various configurations and are commonly influenced by the proximity of the human and the robot [5]. Several innovation for the soundness of more established people could empower them to live autonomously, socially drew in, and safely[6]. Innovation has radically changed the structure and association of nursing to such an extent that the modular changes in present day social insurance have changed nursing's conveyance of care[7]. The individual focused consideration in nursing is guided by the way towards "realizing people as mindful in nursing". The aim of the competency in trends of technology as Caring in Nursing can be actualized at various levels, concentrated on the individual, and complete as opposed to correcting the individual or supplanting the individual's lost or required parts [8]. The consistent sharing of data and incorporated correspondence is fundamental to the accomplishment of individual focused consideration. This joining and sharing can be finished through mechanical methods [9].The ongoing improvements in IoT innovation can assume a huge job in the plan of a reasonable human services framework for more established people. Shrewd homes can change the technique wherein medical services are being given to more seasoned people by coordinating well being[10]. IoT may take further more deep into it by simply advising the person under ailment of their status in Medical Progress, associated gadget innovations can give a start to finish clinical arrangement. A model is savvy bundling. Sensors in an associated medicine bundle can send data from when a bundle is opened, which will be transmitted through a safe system stage for the board which permits checking [11].There is an inclination to present high revolution and apply in care. Assisted living (AL) is the quickest emergent type of more seasoned grown-up people. Robots could reinforce the various kinds of feeling of self-rule in different manners.Robots may reinforce the feeling of self-rule of more seasoned people in AL. They may at the same time represent a danger. Multi-proficient conversations are required on whether robots are invited in care, and on the off chance that they are, the manner by which, for whom, and in what areas[12]. The likelihood that humanoid nurture robots (HNRs) can be conceivable. Being human in relationship with falsely keen machines won't just test the utilitarian parts of savvy machines, yet it will likewise be material to their connections as machines to human beings[13].The general innovative work may get quicken by utilizing a typical robot stage among a great deal of analysts since that empowers them to share their examination results. Accordingly, we have built up a minimal and safe research stage, Human Support Robot (HSR), which can be worked in a real home condition. The

IOP Conf. Series: Materials Science and Engineering 1049 (2021) 012013 doi:10.1088/1757-899X/1049/1/012013

robot rivalry will be effective primarily to actuate the network[14]. The focal point of our exploration was to research whether a social robot can assume control over a portion of this information assortment by controlling surveys autonomously[15]. The technique takes care of the accompanying three concurrent issues: the geometric parameter recognizable proof that transforms the human model to the robot model, movement getting ready for a robot, and the backwards kinematics of the human movement catch information. Synchronous arrangements can impersonate the first movement more precisely than ordinary methodologies, which take care of the issues sequentially[16]. Through support learning with human criticism, the robot is empowered to settle on proper choices by investigating its own inspiration, condition upgrades and human information sources, permitting the robot to complete ambiguous assignments, for example to serve, help or have a discussion with human on various circumstances. In our test, the human fulfillment rate could arrive at 94% and the robot could also keep up a steady condition to perform long haul service[17]. A novel way to deal with coordinating people and assistive robots into ECHONET based savvy home conditions through a multi-modular UI, toward moderating nursing care issues in maturing societies[18]. The ECHONET based brilliant home condition, iHouse, with the all-inclusive AAL arrangement gives a proficient method to improve the personal satisfaction of keen home inhabitants. We have affirmed that a humanoid robot friend can access the iHouse arrange and give client mentioned information and administrations through verbal connection with the help of a characteristic language preparing tool[19]. In the light of segment change and absence of nursing staff, telerobotic frameworks are a promising arrangement for example for homecare in provincial territories or in isolate stations for dispersal of pathogens[20].

3. Proposed Method Akm Med Assistive Bot

The AKM Med Assistive Bot is based on the 'Line Following robot' concept. There would be a Line following tape which is sticked to the floor. The robot traces its path based on the locations where the line following tape is fixed. An Infra-red sensor module is fixed at the front of the robot so that it can trace its path. A pair of ultrasonic sensor is placed at the front and back of the robot system. The Ultrasonic sensor is placed, so that if an obstacle is detected by the ultrasonic sensor, the robot system may stop moving further. The RFID module is intended for authentication and for information gathering purposes. All the movements and actions are commanded through the arduino microcontroller.

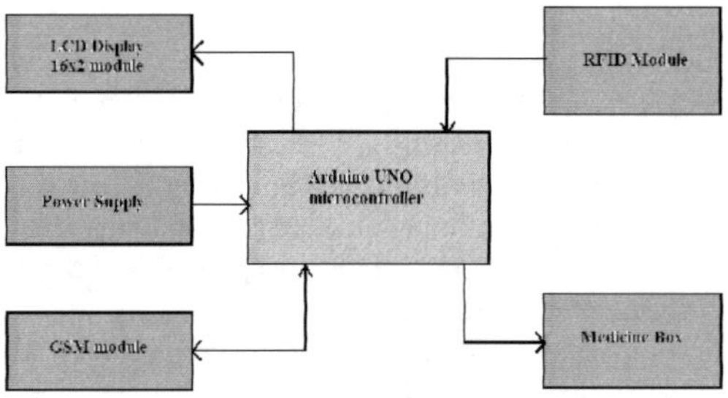

Figure 3.1. Functional Block Diagram

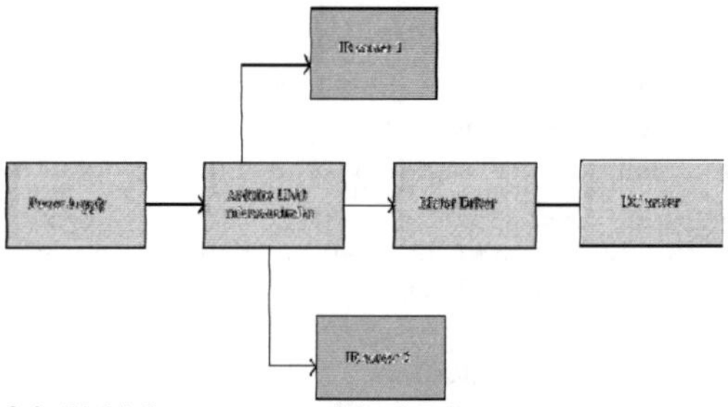

Figure 3.2. Vehicle movement Mechanism

3.1 Working Principle Of Akm Med Assistive Bot

The AKM Med Assistive Bot is equipped with an IR sensor module at the front of the electric vehicle. Two Ultrasonic sensors are placed in front and in back of the vehicle to detect any obstacle and stop accordingly depicted in figure3.1 and 3.2. A line following tape is fixed prior on the floor to aid the movement of the robot and trace its path. If any obstacle is detected by the ultrasonic sensor, the robot automatically stops further movement. Also, a white tape is fixed at the desired location or near thetient to stop the movement of the Med Assistive Bot. The time at which the robot should make its movement to the patient is programmed using the timer module and is interfaced with the arduino microcontroller board. The RFID module is used for authentication. The patient holds the RFID key and card. The RFID key holds the patient's information and the card hold the medicinal data of the patient. The amount of medicine to be taken by the patient is then indicated to the patient and the patient uses the medicine box to take the required quantity of medicines. If in case the number of medicines are above 2 or else if the patient is uneducated, the proposed system can label the medicine as numerical digits, so that the patient can consume the specific labeled medicines.

3.2 Flow Diagram

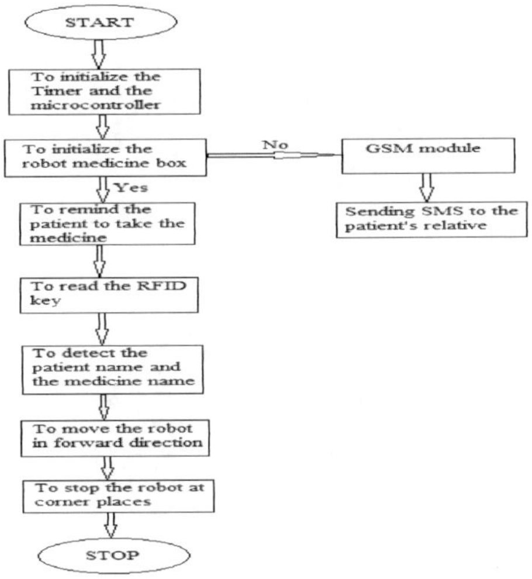

Figure 3.3. Flow Diagram

Flow Diagram Description

***Initialization of the Robot*:** The internal components of the robot system are subjected to initialization for the proper working of all the electronic components.

***Movement towards the patient*:** After initialization, the Infrared sensors sense the presence of the line following tape. The robot traces the path using the infrared sensors by keeping the line following tape as a reference.

***Obstacle detection*:** A pair of ultrasonic sensors are attached to detect the presence of any obstacle during vehicle transit. If an obstacle is detected, the vehicle is stopped and waits until the way is clear.

Authentication: An authentication step is done so as to confirm whether the medicines are taken by the right patient. The authentication is carried by the RFID module. The RFID module holds information about the patient along with the medicines to be taken by them.

Recommendation: If the authentication is successful, the system reminds the patient to consume the required medicine. The robot system stands near the patient area for a while until the patient takes the medicine. If no activity if taken by the patient, an alarming system activates alerting the nearby nurse or doctor for further help. If the medicine is taken successfully, the status of the medication is sent to the patient"s relative through SMS. The GSM module sends the status of the medicine to the patients relative mobile by means of an SMS.

4. Results And Discussion

We have developed this prototype so as to help the patients in taking their daily medications without the actual need of a human nurse or a care taker. As for now, It acts as a semi-autonomous system, meaning at the start of each day, a human nurse is required in order to feed the information of the medicines that will be taken by the patient each day. This can help the nurses and other care takers to focus on their tasks effectively.

Figure 4.1. AKM Med Assistive Bot Prototype

Simulation Results

PROTEUS DESIGN SUITE VERSION 8.0 is the software tool used to build the project in software simulation. In simulation, the components are taken which are included in the hardware part.
First, the Arduino coding developed for running the whole circuit will be dumped into the micro controller. The Coding for the Infrared Sensor was built and it be dumped into the circuit for detecting the patient. After dumping the code into the components, the circuit will be operated by clicking on the *RUN* Button. After running the circuit, in the simulation output figure.4.2.shows the *patient name and the respective medications* it displayed in the LCD display Module. At this point, the motor driver circuit will be working according to the logics of 00, 01, 10 and 11for the movements of the robot.After successful authentication, the robot will display the Patient Name, Medications to be Taken, as in figure.4.3. the *Status of the Medications* will be displayed. The Patient is to take the medicine which is to be taken, and its status will be shown when the patient click on the button switch. The Button Switch is used in the Circuit because of the in availability of RF module in the proteus design suite software.After taking the medications, the patient has to click on the Switch button, so that it will indicate that the patient has taken the medicine it is showed in figure.4.3 like *medication taken.*The Motor Driver Circuit will be controlling the movement of the robot with the help of 4 motors.

IOP Conf. Series: Materials Science and Engineering 1049 (2021) 012013 doi:10.1088/1757-899X/1049/1/012013

S.NO	CODE	MOVEMENTS
1	01	RIGHT
2	10	LEFT
3	11	FORWARD
4	00	STOP

Table.1. Coding for Robots movement

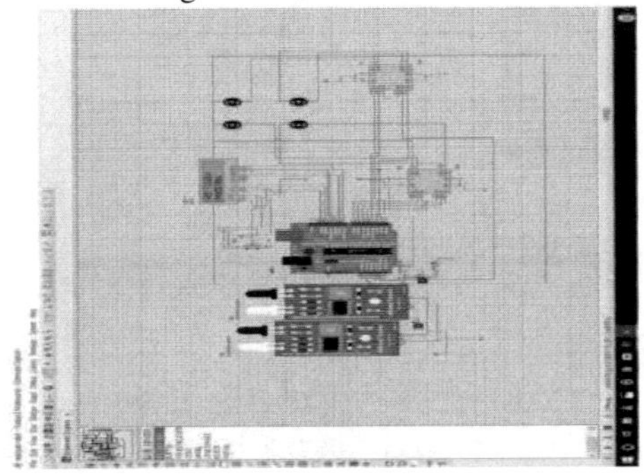

Figure 4.2. Simulation Output 1

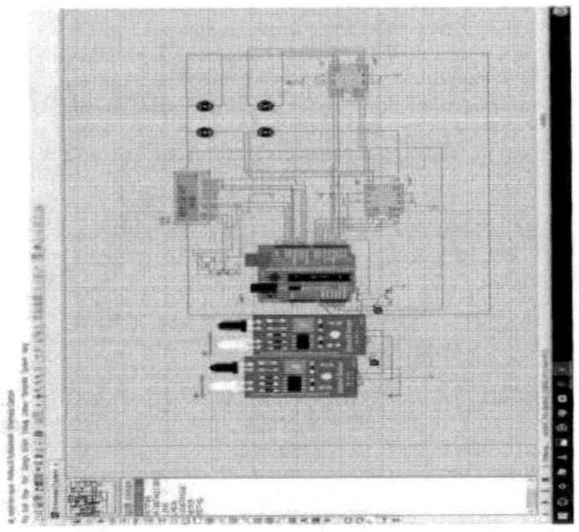

Figure 4.3. Simulation Output 2

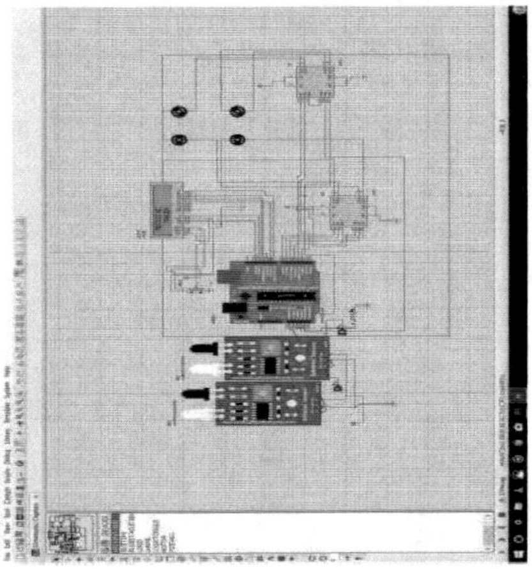

Figure 4.4. Simulation Output 3

5. Conclusion And Future Work

The AKM Med Assistive bot helps the patients in hospitals and other old age homes to periodically consume the required amount of tablets without the intervention of a human nurse. The robot system recommends the amount of medicine that may be taken by the patient at a particular time. After taken the medication the message will be given to the patient relatives. Currently, the system is designed for one individual patient, and in future, this work have planned to implement the nursing robot system for more than one individual. In Future, it will make the system to work as a fully autonomous system. It also embed other hardware components so that various parameters such as the heart rate, blood sugar level, activity tracking, can also be fed into the system and be monitored periodically.

References

[1] Jiang, C., & Ueno, S. (2019). Posture maintenance control of 2-link object by nonprehensile two-cooperative-arm robot without compensating friction. *IEEE/CAA Journal of Automatica Sinica*, *6*(6), 1397-1403.

[2] Glauser, W. (2017). Artificial intelligence, automation and the future of nursing. *The Canadian nurse*, *113*(3), 24-26.

[3] Tanioka, T., Yasuhara, Y., Osaka, K., Ito, H., & Locsin, R. (2017). Supervising Editor: Nursing Robots: Robotic Technology and Human Caring for the Elderly.

[4] Ahn, H. S., Choi, J., Moon, H., & Lim, Y. (2018, March). Social human-robot interaction of human-care service robots. In *Companion of the 2018 ACM/IEEE International Conference on Human-Robot Interaction* (pp. 385-386).

[5] van Greunen, D. (2019, February). User Experience for Social Human-Robot Interactions. In *2019 Amity International Conference on Artificial Intelligence (AICAI)* (pp. 32-36). IEEE.

[6] Pepito, J. A., Locsin, R. C., & Constantino, R. E. (2019). Caring for older persons in a technologically advanced nursing future. *Health*, *11*(5), 439-463.

[7] Rouleau, G., Gagnon, M. P., Côté, J., Payne-Gagnon, J., Hudson, E., & Dubois, C. A. (2017). Impact of information and communication technologies on nursing care: Results of an overview of systematic reviews. *Journal of medical Internet research*, *19*(4), e122.

[8] Pepito, J. A., & Locsin, R. (2019). Can nurses remain relevant in a technologically advanced future?. *International journal of nursing sciences*, *6*(1), 106-110.

[9] American Geriatrics Society Expert Panel on Person-Centered Care, Brummel-Smith, K., Butler, D., Frieder, M., Gibbs, N., Henry, M., ... & Saliba, D. (2016). Person-centered care: A definition and essential elements. *Journal of the American Geriatrics Society*, *64*(1), 15-18.

[10] Majumder, S., Aghayi, E., Noferesti, M., Memarzadeh-Tehran, H., Mondal, T., Pang, Z., & Deen, M. J. (2017). Smart homes for elderly healthcare—Recent advances and research challenges. *Sensors*, *17*(11), 2496.

[11] Treskes, R. W., Van der Velde, E. T., Schoones, J. W., & Schalij, M. J. (2018). Implementation of smart technology to improve medication adherence in patients with cardiovascular disease: is it effective?. *Expert review of medical devices*, *15*(2), 119-126.

[12] Pirhonen, J., Melkas, H., Laitinen, A., & Pekkarinen, S. (2020). Could robots strengthen the sense of autonomy of older people residing in assisted living facilities?—A future-oriented study. *Ethics and Information Technology*, *22*(2), 151-162.

[13] Rozzano C. Locsin, Hirokazu Ito, Tetsuya Tanioka, Yuko Yasuhara, Kyoko Osaka, Savina O. Schoenhofer, Humanoid Nurse Robots as Caring Entities: A Revolutionary Probability? International Journal of Studies in Nursing; Vol. 3, No. 2; 2018.

[14] Yamamoto, T., Terada, K., Ochiai, A., Saito, F., Asahara, Y., & Murase, K. (2019). Development of Human Support Robot as the research platform of a domestic mobile manipulator. *ROBOMECH journal*, *6*(1), 4.

[15] Van der Putte, D., Boumans, R., Neerincx, M., Rikkert, M. O., & de Mul, M. (2019, March). A social robot for autonomous health data acquisition among hospitalized patients: an exploratory field study. In *2019 14th ACM/IEEE International Conference on Human-Robot Interaction (HRI)* (pp. 658-659). IEEE.

[16] Ayusawa, K., & Yoshida, E. (2017). Motion retargeting for humanoid robots based on simultaneous morphing parameter identification and motion optimization. *IEEE Transactions on Robotics*, *33*(6), 1343-1357.

[17] Yang, C. Y., Lu, M. J., Tseng, S. H., & Fu, L. C. (2017, September). A companion robot for daily care of elders based on homeostasis. In *2017 56th Annual Conference of the Society of Instrument and Control Engineers of Japan (SICE)* (pp. 1401-1406). IEEE.

[18] Bui, H. D., & Chong, N. Y. (2018, September). An integrated approach to human-robot-smart environment interaction interface for ambient assisted living. In *2018 IEEE Workshop on Advanced Robotics and its Social Impacts (ARSO)* (pp. 32-37). IEEE.

[19] Bui, H. D., Pham, C., Lim, Y., Tan, Y., & Chong, N. Y. (2017, November). Integrating a humanoid robot into echonet-based smart home environments. In *International Conference on Social Robotics* (pp. 314-323). Springer, Cham.

[20] Panzirsch, M., Weber, B., Rubio, L., Coloma, S., Ferre, M., & Artigas, J. (2017, June). Tele-healthcare with humanoid robots: A user study on the evaluation of force feedback effects. In *2017 IEEE World Haptics Conference (WHC)* (pp. 245-250). IEEE.

IOP Conf. Series: Materials Science and Engineering 1049 (2021) 012014 doi:10.1088/1757-899X/1049/1/012014

Classification of Pituitary Tumor and Multiple Sclerosis Brain Lesions through Convolutional Neural Networks

Shrwan Ram
Anil Gupta

Department of Computer Science and Engineering
MBM Engineering College,
Jai Narain Vyas University,
Jodhpur, Rajasthan, INDIA

Abstract: Automatic classification of Brain Tumor and brain Lesions has become a very important step in the field of medical image analytics. The machine learning/Deep learning approaches are playing a tremendous role in the field of medical imaging classification, due to the drastic changes in the field of computing power and image analytics techniques. The deep learning, which is the subfield of machine learning, is playing the major role in the automatic classification of Magnetic Resonance Images (MRIs) having various brain abnormalities. Convolutional Neural Networks are widely used for the classification and detection of various brain disorders. In this research paper, Convolutional Neural Networks are designed with considering various learning parameters for the classification of Multiple Sclerosis Brain Lesions and Pituitary Tumor. In the proposed research, T1-weighted Contrast-enhanced Magnetic Resonance images are preprocessed with various image-preprocessing approaches such as to resize the images, to convert the images into suitable image format so that the experimental work can be performed with deep learning in the Matlab environment. The Experiment is conducted with the dataset of Multiple Sclerosis and Pituitary Tumor each of having 718 and 930T1-weighted MRI images respectively. The experimental results we achieved 99.7% classification accuracy of pituitary Tumor, and 99.2% accuracy of Multiple Sclerosis brain Lesions. The average accuracy of both classifications is 99.55%. The precision of the classification of Pituitary Tumor is 99.7, recall value is 99.7 and the f1_score of the classification is 99.7%. Similarly, the Precision of the classification of Multiple Sclerosis Brain Lesions is 99.15%, the recall value is 99.15%, and the f1_score is 99.15%. The purposed approach of the Convolutional Neural Network architecture exhibited outstanding performance as compared to other research outcomes.

Keywords- Brain Tumor, Multiple Sclerosis Brain Lesions, machine learning, Deep learning, Convolutional Neural Networks, brain disorders.

1 INTRODUCTION

Brain Tumor and Multiple Sclerosis both are considered very dangerous diseases related to the brain [1]. The Pituitary tumor inside the pituitary gland affects the functioning of the brain. The pituitary gland is a very important organ in the human body; it plays a central role in the growth of body, metabolism, and reproductive function. The pituitary Tumors patients are increasing very rapidly all across the world. More than 14,120 patients are diagnosed in the United States, last one year as reported by the World Health Organization [2]. A pituitary tumor is generally benign but causes other mental damage as compared to other brain tumors [3]. Multiple sclerosis (MS) is a leading disabling disease related to the brain and spinal cord (Central Nervous System). In this disease, the human immune system attacks the myelin sheath, a protective cover for nerve fibers. The damage due to Multiple sclerosis causes communication problems between the human brain and other body parts. The disease can cause permanent damage to the nerves. Early detection of the disease can help for improvement [5]. The advancement of the latest technologies such as machine learning and deep learning are playing a major role in analyzing the large medical imaging datasets by providing the important supporting tools for the classification and prediction of various brain abnormalities [3]. The Convolutional Neural network is one of the machine learning approaches that substantially performed outstanding results in the classification of medical images [3]. The various machine-learning methods for image classification and disease detection are applied in MRI image analytics to provide additional support and opinions to radiologists [3].

In this research work, the "Convolutional Neural Network" [15, 16, 17], model is proposed for the classification of Pituitary Tumor and Multiple sclerosis brain lesions. The model is built and trained considering the various hyper-parameters. The size of the input images, size of the image dataset, and the number of categories of the images are very important parameters. The Convolutional Neural Network is trained with setting Hyper-parameters such as learning rate, dropout factor, number of Convolutional layers, etc. The paper is organized into five sections such as introduction, literature review, proposed research methodology, experimental results, and the conclusion with future work. In the literature review section, related research work carried out by the academicians, and researchers have explored with research the outcomes. In the research methodology, proposed research approaches are explored with considering the concepts of Convolutional Neural Networks (CNNs) and image preprocessing techniques. An experimental setup with detailed implementation approaches is explored in the results and analysis section. Finally, the conclusion section highlights the research outcome with future research directions.

2 RELATED WORK

Many researchers and academicians are continuously working in the field of medical image analytics. Different types of methodologies and models are developed and deployed for the classification and detection of various brain disorders. Through the depth literature reviews, I came across the various research papers related to disease diagnosis through Convolutional neural networks using Magnetic Resonance Images (MRIS). The following research papers are more relevant to the proposed research work:

Sunanda Das et al. [1] proposed the Convolutional Neural Network (CNN) for the classification of different types of brain tumors, such as Glioma Tumor, Meningioma Tumor, and Pituitary Tumor. The researchers have performed the task of brain tumor classification using 3064 T1 weighted contrast-enhanced MRI images. The CNN model was trained using a series of Convolutional and pooling operations. The MRI images are resized and Convolutional operations are performed with Convolutional filters/kernel of variable size to generate a feature map. Accuracy of up to 94.39% was achieved after evaluating the model on test datasets with 28.16% loss. The precision of 88%, 94%, and 98% was achieved for Glioma Tumor, Meningioma Tumor, and Pituitary Tumor respectively.

Melika Maleki, M. Teshnehlab, and M. Nabavi [5] developed the fully automated process of extracting features from magnetic resonance images (MRI) of multiple sclerosis. Based on the

extracted features, multiple sclerosis disease was classified. The authors proposed the Convolutional Neural Network for the extraction of features from MRI images and the multilayer neural network for the classification of two classes such as normal and Multiple sclerosis. The outcomes of the research shown that CNNs approaches for the classification are more efficient and accurate without using the image segmentation approach. The authors achieved the classification accuracy as 92.93, sensitivity as 96.12, and specificity as 97.57.
Mostafa Salem et al. [6] proposed the fully Convolutional Neural Network for the detection of multiple sclerosis lesions in T2-w MRI images. The author's proposed network was trained using four image modalities such as T1-w, T2-w, PD-w, and Fluid Attenuated Inversion Recovery (FLAIR) with a gradient descent optimization approach. The authors trained the network with 3-dimensional 32X32X32 patches with a step size of 16x16x16 were extracted from follow-up images and baseline images of the four input modalities. The researchers detected the new lesions with a mean Dice Similarity coefficient value as 83%, the true positive rate for the detection as 83.09% and, false-positive detection rate as 9.36%. Further, they found that larger patch sizes did not significantly improve the performance. Increasing the size of patches requires more computational power and processing memory.

3 RESEARCH METHODOLOGY

This part of the research paper focused on the research methodology that is used to classify the Pituitary Tumor and Multiple Sclerosis Brain Lesions. The classification is performed with the following steps: The most important part of the research is the suitable data collection for practical implementation. Choosing a suitable dataset decides the performance of the implemented modal. The second step is the image preprocessing used for enhancing the quality of the raw images data by performing various data transformation techniques. The third step of the methodology is the selection of Convolutional Neural Network architecture with suitable hyper-parameters for the efficient execution accurate classification of the Brain Tumor and MS lesions. The research methodology is based on the design and implementation of a Convolutional Neural Network suitable for classifying the Magnetic Resonance Images with considering the various learning parameters. Various CNNs architectures were proposed by the researchers with classification accuracies, not more than 95 or 96%.

A. *Data Collection and Description of Data*

The Pituitary tumor dataset collected from [8] contains 930 T1-weighted contrast-enhanced images. The Pituitary images are collected from 233 patients. The Multiple sclerosis images are retrieved from [11], [12], [13], [14] contains 38 patients MRI images in TIFF format. The multiple sclerosis images are the collection of first and the second examination in 0 months and 6-12 months. Other images are retrieved online and combined with the Multiple Sclerosis images.

B. *Preprocessing*

Both the datasets are resized as 512x512-image size and converted into PNG image file format. The pituitary tumor dataset originally downloaded in the .mat file format is converted into.PNG image file format. The datasets are kept in two different folders such as Pituitary and multiple sclerosis. The Pituitary and Multiple sclerosis images are shown in figure 1 and figure 2.

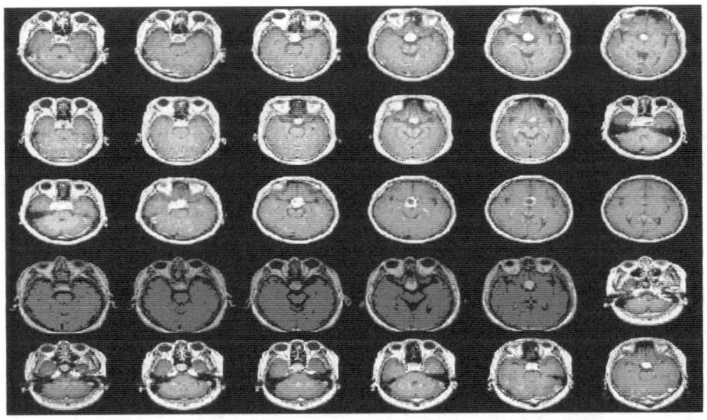

Figure 1. Pituitary Tumor Sample images [8].

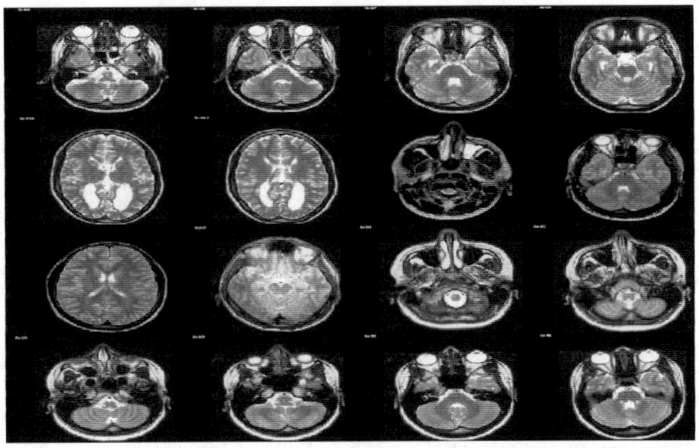

Figure. 2. Multiple Sclerosis Sample Images [11], [12], [13], [14]

The datasets are divided into training and validation datasets containing 75% data for training and 25% data for validation. The total image data file are (914+718) =1632 images. 1200 image files are used for training the Convolutional Neural Network model and the remaining 432 files are used for the validation of the model.

C. *Pituitary Tumor and Multiple Sclerosis Brain Lesions Classification using CNN.*

The Convolutional Neural Networks are trained to learn a set of features from the input images that can be used to classify new images data. Convolutional Neural Networks performance is tested in the field of brain MRI image analytics, specifically brain tumor/lesions classification and detection [6]. Convolutional Neural Network comprises the ability to extract and learning the features from large image datasets through the training process with gradient descent and computer vision approaches as compared to simple Artificial Neural Network. Through the design of Convolutional Neural Network layered architecture a series of Convolutional, batch normalization, ReLU, and max-pooling operations, followed by a fully connected layer, Softmax layer, and classification layer [1].

In this research work, a Convolutional Neural Network is proposed with image input size 512x512. 8 filters/kernels of size 11x11 are used in the first Convolutional layer with stride [1 1] and padding [1 1 1 1]. The process of stride slides the movement of the filter at each step horizontally as well as vertically. The proposed Convolutional Neural Network is designed with three Convolutional layers, Three ReLU layers, and three batch normalization layers and, three Max-pooling.

All the layers are finally connected with a fully connected layer, two-way softmax layer, and classification layer. The ReLU activation function is used for transforming the weighted input from the input node to the next output node as a positive input. The Rectified Linear Unit activation function is a piecewise linear function that is used to eliminate the negative values after Convolutional operation. This function has become the default activation function for many types of Convolutional Neural Networks. It is shown through the implementation of CNN that the ReLU activation function achieves comparatively better Performance.

The Rectified Linear Unit function can be defined as [15]:

$$f_x = max(0, wx + bias) \quad (1)$$

Where w is the weight and x as an input with bias value.

The Convolutional operation and max-pooling operation both are performed by CNN using the following two mathematical formulas:

$$\text{Convolution-Layer-Output} = (W - F + 2P)/S + 1 \quad (2)$$

$$Maxpooling - Output = (W - F_m + 2P)/S_m + 1 \quad (3)$$

Where W is the width of the image, the filter size is F and the stride is S and zero paddings are P. we can choose either image height or width if the height and width are equal, otherwise, the calculation is performed with image height and width separately. In the case of Max-pooling, the filter size is F_m and stride is S_m as shown in equation 3. The image size in the proposed work is 512x512, so the image width and height both are equal. In this research paper, we took image size as 512x512, filter size as 11x11 with zero paddings as [1 1 1 1], and stride is [1 1] during the first Convolutional operation, the output of the Convolutional layer is ((512-11+2))/1+1, that is equal to 504, so we get 504x504x8 feature maps in the first Convolutional process. After the Convolutional and other operations, max-pooling is performed on the feature maps collected during the Convolutional process. Formula number *(2)* can be used for that purpose. Therefore, the output after the first max-pooling operation is:

((504-2+0))/2+1, that is equal to 252, so we obtained 252x252x8 feature maps after the first max-pooling operation. The rest of the Convolutional and max-pooling operations are performed using the formula number *(2)* and *(3)*.

The resulting output after the first Convolutional operation is 504x504x8. Batch Normalization operation is performed on the resulting output of the first Convolutional operation. This operation normalized the output within the range. ReLU Operation is performed to eliminate the negative values. The max-pooling operation with filter size 2x2 with stride [2 2] is performed on the first convolved output and the final output is generated as 252x252x8. In the second Convolutional operation, 16 filters of size 11x11 are applied to the input image size 252x252 to obtained the resulting image size is 244x244 with weights as 11x11x8x16 and bias as 1x1x16. The batch normalization and ReLU operation are also performed on the resulting output. Through the second max-pooling operation down-sampling is performed with a 2x2 filter size with [2 2] as stride. The resulting feature maps of size 122x122x16 are obtained. Through the third Convolutional operation with 32 filters/kernel of size 11x11, resulting feature maps of size 114x114x32 are generated with 11x11x16x32 weights and 1x1x32 Bias.

Through the third down-sampling process with2x2 filter and [2 2] stride, feature maps are further reduced and resized as 57x57x32. Finally, 1x1x2 actions with 2x103968 weights are connected to a fully connected layer. The fully connected layer is further connected to softmax and classification to classify the Pituitary Tumor and Multiple Sclerosis brain lesions in the form of two classes.

IOP Conf. Series: Materials Science and Engineering 1049 (2021) 012014 doi:10.1088/1757-899X/1049/1/012014

D. *Performance Evaluation of the Proposed CNN architecture.*

The performance of the proposed Convolutional Neural Network for the classification of Pituitary Tumor and Multiple Sclerosis brain lesions is evaluated based on classification accuracy, Precision, Recall, and f1_score.

$$Accuracy = \frac{(True\ Positive\ +\ True\ Negative)}{(True\ Positive +\ False\ Positive\ +\ False\ Negative\ + True\ Negative)} \quad (4)$$

$$Precision = \frac{True\ Positive}{(True\ Positive + False\ Positive)} \quad (5)$$

$$Recall = \frac{TruePositive}{(TruePositive + FalseNagetive)} \quad (6)$$

$$f\_score = \frac{2 * \text{True Positive}}{(2 * TruePositive + FalsePositive + FalseNegative)} \quad (7)$$

4 EXPERIMENTAL SETUP AND RESULTS ANALYSIS

The model is implemented in the Matlab Environment using Matlab version R2020a with the help of Deep learning and computer vision approaches [18]. The Stochastic Gradient Descent Momentum (SGDM) optimizer, 128 epochs, and 30 iterations for validation frequency are used as a hyper-parameter during the training. SGDM is one of the powerful optimization algorithms generally used to train the Convolutional Neural Networks. The model is trained on Quadro P5000 GPU with 16GB GDDR5X GPU Memory, 2560 NVIDIA CUDA Cores, and 8.9 Teraflops Computing Power. Figure 3 shown below illustrates the training progress with training accuracy, validation accuracy, training loss, and validation loss. The outcomes of the Convolutional Neural Network during the training period are shown below in Table 1.

Table 1 Convolutional Neural Networks. Layer wise Training outcomes

	Name	Type	Activations	Learnable
1.	Input 512x512x1 images with Zero-center normalization	Image Input	512x512x1	---
2.	Convlayer1 8x11x11x1 Convolutional with stride [1 1] and padding [1 1 1 1]	Convolutional	504x504x8	Weights 11x11x1x8 Bias 1x1x8
3.	BatchNorm1 Batch normalization with 8 channels	Batch Normalization	504x504x8	Offset 1x1x8 Scale 1x1x8
4.	Relu1 ReLU	ReLU	504x504x8	--
5.	Maxpool1 2X2 max pooling with stride [2 2] and padding [0 0 0 0].	Max Pooling	252x252x8	--
6.	Convlayer2 16 11x11x8 Convolutional with stride [1 1] and Padding [0 0 0 0].	Convolutional	244x244x16	Weights 11x11x8x16 Bias 1x1x16
7	BatchNorm2	Batch Norma-	244x244x16	Offset

.	Batch normalization with 16 channels	lization		1x1x16 Scale 1x1x16
8 .	Relu2 ReLU	ReLU	244x244x16	--
9 .	Maxpool2 2X2 max pooling with stride [2 2] and padding [0 0 0 0	Max Pooling	122x122x16	--
1 0.	Convlayer3 32 11x11x16 Convolutional with stride [1 1] and Padding [1 1 1 1].	Convolutional	114x114x32	Weights 11x11x16x32 Bias 1x1x32
1 1.	BatchNorm3 Batch normalization with 32 channels	Batch Normalization	114x114x32	Offset 1x1x32 Scale 1x1x32
1 2.	Relu3 ReLU	ReLU	114x114x32	--
1 3.	Maxpool3 2X2 max pooling with stride [2 2] and padding [0 0 0 0]	Max Pooling	57x57x32	---
1 4.	2 Fully Connected Layer	Fully Connected	1x1x2	Weights 2x103968 Bias 2x1
1 5.	SoftmaxLayer softmax	softmax	1x1x2	---
1 6.	Classification layer Cross-entropy with classes 'Msclerosis' and 'pituitary'	Classification Output.	-	-

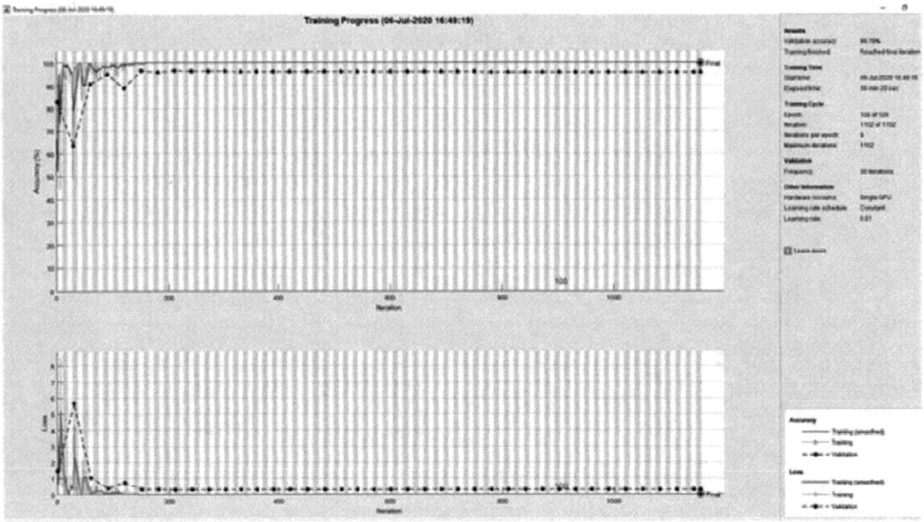

Figure 3. Execution Progress with training accuracy, validation accuracy, training loss, and validation loss

The confusion matrix is an important performance evaluation matrix used to depict the predicted classification results concerning actual values. Through the validation process, the confusion matrix is generated with predicted and actual classification results in the form of the Confusion matrix as shown in figure 4. During the validation, 118 images of Multiple Sclerosis and 330 images of Pituitary Tumor are used for validation. We can that the proposed Convolutional Neural Network truly predicted 117 images belong to Multiple Sclerosis and 329 images are predicted as Pituitary Tumor. The validation process also creates a confusion matrix chart showing the true labels and predicted labels with predicted accuracy and miss classification accuracy. The rows of the confusion matrix depict true classes and columns depict the predicted class. The main diagonal and opposite diagonal cells represent correctly classified and incorrectly classified observations respectively [18].

A set of 53 images of Pituitary Tumor and Multiple Sclerosis are used for the testing of the classification model. The testing accuracy is printed above each image. Among the 50 testing images, some images with testing accuracy are printed above on each image, are shown in figure 6.

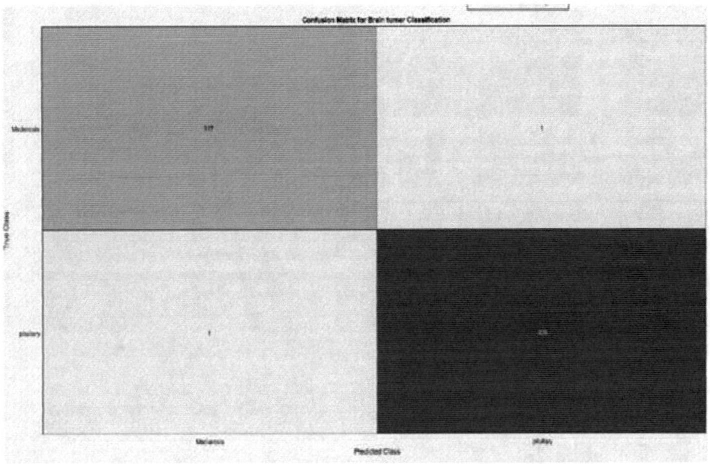

Figure 4. Confusion Matrix for the Classification.

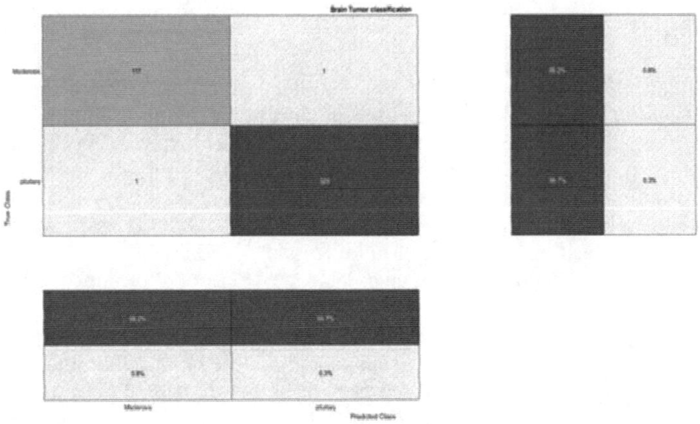

Figure 5. Confusion Chart for the Classification

Table 2 Performance Evaluation Table

Class	Accuracy	Precision	Recall	f1_score
Pituitary Tumor	0.9955	0.9970	0.9970	0.9970
Multiple Sclerosis Brain Lesions	0.9955	0.9915	0.9915	0.9915

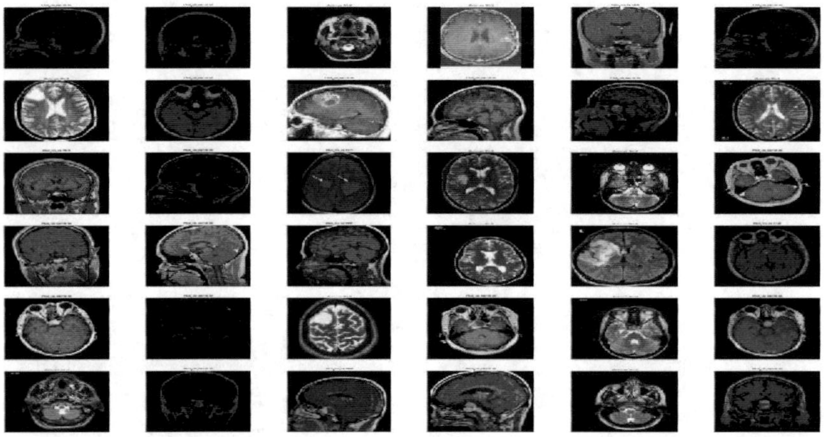
Figure 6. Classification results with an accuracy of classification printed above the image

We can visualize all the images as shown in figure 6 by the zooming option. The network for classification of both datasets is obtained through the Convolutional Neural network training process. The classification accuracy is printed above each image during the testing of the network.

5 CONCLUSIONS

Accurately classifying brain disorders is a challenging task for healthcare experts. Deep learning approaches are proving to be effective and accurate methods of automatic brain tumors and lesions classification and detection. This research paper explored the deep learning approaches with the help of the Convolutional Neural Network to provide more accurate results for the classification of brain-related diseases. The 99.2% classification accuracy is achieved for the Multiple Sclerosis brain lesion classification and 99.7% classification accuracy is achieved for the classification of Pituitary Tumor. It is also found that classification accuracy is highly dependent on the quality of the image datasets. The model designed through the CNN approaches can be used to classify the unknown images of the tumors and lesions if the images are scanned with a high-quality MRI machine of 1.5 Tesla, 3 Tesla or higher intensities. As compared to the other classification model, our model for the brain tumor and Multiple Sclerosis brain lesion, classification has higher accuracy as shown in table 2. The model can be used for real-life applications if we have enough MRI images data related to Pituitary Tumor and Multiple Sclerosis. As a future scope, we can explore the research work in the domain of medical image processing applications for the detection of various brain-related disorders through transfer learning approaches as well as designing the large Convolutional Neural Networks.

REFERENCES

[1] Sunanda Das, O.F.M. Riaz Rahman Aranya, and Nishat Nayla Labiba, "Brain Tumor Classification Using Convolutional Neural Network', 1st International Conference on Advances in Science, Engineering and Robotics Technology 2019 (ICASERT 2019).

[2] Pituitary Gland Tumor: Statistics, Available [online]. https://www.cancer.net/cancer-types/pituitary-gland-tumor/statistics.

[3] Milica M. Badza and Mark C. Barjaktarovic, "Classification of Brain Tumors from MRI images using a Convolutional Neural Network.

[4] Katz Sand I. Classification, diagnosis, and differential diagnosis of multiple sclerosis. Curr Opin Neurol. 2015; 28(3):193-205. doi:10.1097/WCO.0000000000000206.

[5] Melika Maleki, Pro M. Teshnehlab, and Dr. M. Nabavi, "Diagnosis of Multiple Sclerosis (MS) Using Convolutional Neural Network (CNN) from MRIs", Global Journal of Medicinal Plant Research, 1(1): 50-54., 2012 ISSN 2074-0883.

[6] Mostafa Salem, Sergi Valverdea , Mariano Cabezasa , Deborah Paretoc , Arnau Olivera , Joaquim Salvia , Àlex Rovirac and Xavier Lladóa, "A fully Convolutional neural network for new T2-w lesion detection in multiple sclerosis", NeuroImage:Clinical Volume 25, 2020, 102149.

[7] Multiple sclerosis: - Available [online] at https://www.mayoclinic.org/diseases-conditions/multiple-sclerosis/symptoms-causes/syc-20350269.

[8] Cheng,Jun(2017): brain tumor dataset. Figshare. Dataset .https://doi.org/10.6084/m9.figshare.1512427.v5

[9] Imaging of the pituitary: Recent advances: available [online] https://www.ncbi.nlm.nih.gov/pmc/articles/PMC3183511/

[10] Commowick O, Istace A, Kain M, et al. Objective Evaluation of Multiple Sclerosis Lesion Segmentation using a Data Management and Processing Infrastructure. Sci Rep.2018; 8(1):13650. Published 2018 Sep 12. doi:10.1038/s41598-018-31911-7

[11] C.P. Loizou, V. Murray, M.S. Pattichis, I. Seimenis, M. Pantziaris, C.S. Pattichis, ìMulti-scale amplitude modulation-frequency modulation (AM-FM) texture analysis of multiple sclerosis in brain MRI images,î IEEE Trans. Inform. Tech. Biomed., vol. 15, no. 1, pp. 119-129, 2011.

[12] C.P. Loizou, E.C. Kyriacou, I. Seimenis, M. Pantziaris, S. Petroudi, M. Karaolis, C.S. Pattichis, ìBrain white matter lesion classification in multiple sclerosis subjects for the prognosis of future disability, î Intelligent Decision Technologies Journal (IDT), vol. 7, pp. 3-10, 2013.

[13] C.P. Loizou, M. Pantziaris, C.S. Pattichis, I. Seimenis, ìBrain MRI Image normalization in texture analysis of multiple sclerosisî, J. Biomed. Graph. & Comput., vol. 3, no.1, pp. 20-34, 2013.

[14] C.P. Loizou, S. Petroudi, I. Seimenis, M. Pantziaris, C.S. Pattichis, Quantitative texture analysis of brain white matter lesions derived from T2-weighted MR images in MS patients with clinically isolated syndromeî, J. Neuroradiol., accepted.

[15] LeCun Y, Bengio Y, Hinton G (2015) Deep learning. Nature 521:436–444

IOP Conf. Series: Materials Science and Engineering 1049 (2021) 012014 doi:10.1088/1757-899X/1049/1/012014

[16] Krizhevsky, I. Sutskever, and G. Hinton. Imagenet classification with deep convolutional neural networks. In NIPS, 2012.
[17] Y. LeCun, B. Boser, J. S. Denker, D. Henderson, R. E. Howard, W. Hubbard, and L. D. Jackel. Backpropagation applied to handwritten zip code recognition. Neural computation, 1989.
[18] MATLAB R2020a, The MathWorks, Inc., Natick, Massachusetts, United States.

Furnace Flame Tracking for Combustion Quality Estimation in Coal Fired Boilers

[1]K.Sujatha, [1]P. Sinthia, [1]Jada Satish Babu, [2]R.S. Ponmagal [2]J. Veerendrakumar, [2]V. Balajiand [2]V. Srividhya

[1]Professor, EEE Department, Dr. MGR Educational and Research Institute, Chennai.
[1]Associate Professor, Biomedical Engineering, Saveetha Engineering College, Chennai.
[1]Assistant Professor, Dept. of EEE., REVA University, Bangalore.
[2]Associate Professor, CSE Department, SRM Institute of Science and Technology, Kattankolathur.
[2]Research Scholar, EEE Department, Dr. MGR Educational and Research Institute, Chennai.
[2]Professor, EEE Department, JNN Institute of Engineering, Chennai.
[2]Assistant Professor, EEE Department, Meenakshi College of Engineering, Chennai.

drksujatha23@gmail.com

Abstract. This research work focus is on optimization of the combustion quality as the combustion quality plays a vital role in minimizing NOx and CO emissions in the flue gas. Power generated is also another important factor under consideration so as to meet the demand. Flame monitoring in boilers using digital imaging provides replacement for the conventional technique of temperature measurement. The indigenous method can be replaced by incorporating an online PC based system by analyzing the flame images.

Keywords:Combustion quality, Image J, Image Processing, K-mean algorithm, Farthest First algorithm, Weka.

1. INTRODUCTION

Boiler efficiency can be optimized by reducing the operating cost and it is one of challenging task in the power stations. The improvement in combustion quality leads to consistency in power generation and reduction in cost [1-6]. Presently, the ON-OFF conditions of the flame inside the furnace alone are determined. Explosion of the boiler can be prevented at the time of start-up conditions. Unmonitored dumping of coal in the furnace is very dangerous. So a boiler flame surveillance system is incorporated [7]. Conveyors are used to transport coal from the coal yard to the grinding mills. It is then crushed in mills and the unwanted materials associated are removed in the ESP which attracts all the fine iron particles. The advantage of using pulverized coal is that the finely preheated coal is uniformly spread on the hearth of the furnace [8,9]. The tangential firing system consists of 12 lignite and 8 oil burners. Oil burners are used for initial firing at 21m. The coal firing takes place at 23m. Hence the entire firing process is over within 23m and the total height of the boiler is about 91m.

2. PROCEDURE FOR FLAME ACQUISITION

The flame video is recorded along with the readings relating to the parameters as indicated in Table 1. The flowchart for flame acquisition is illustrated in Figure 1.

Table.1 Experimental Data from NLC

S.No.	% O_2 (Left)	% O_2 (right)	% O_2 (Average)	Load (MW)	Temperature of the flame °C
1.	4	3.8	3.9	210	897
2	3.6	4	3.8	210.6	912
3	3.4	3.9	3.65	210.8	913
4	3.2	3.4	3.3	212.21	933

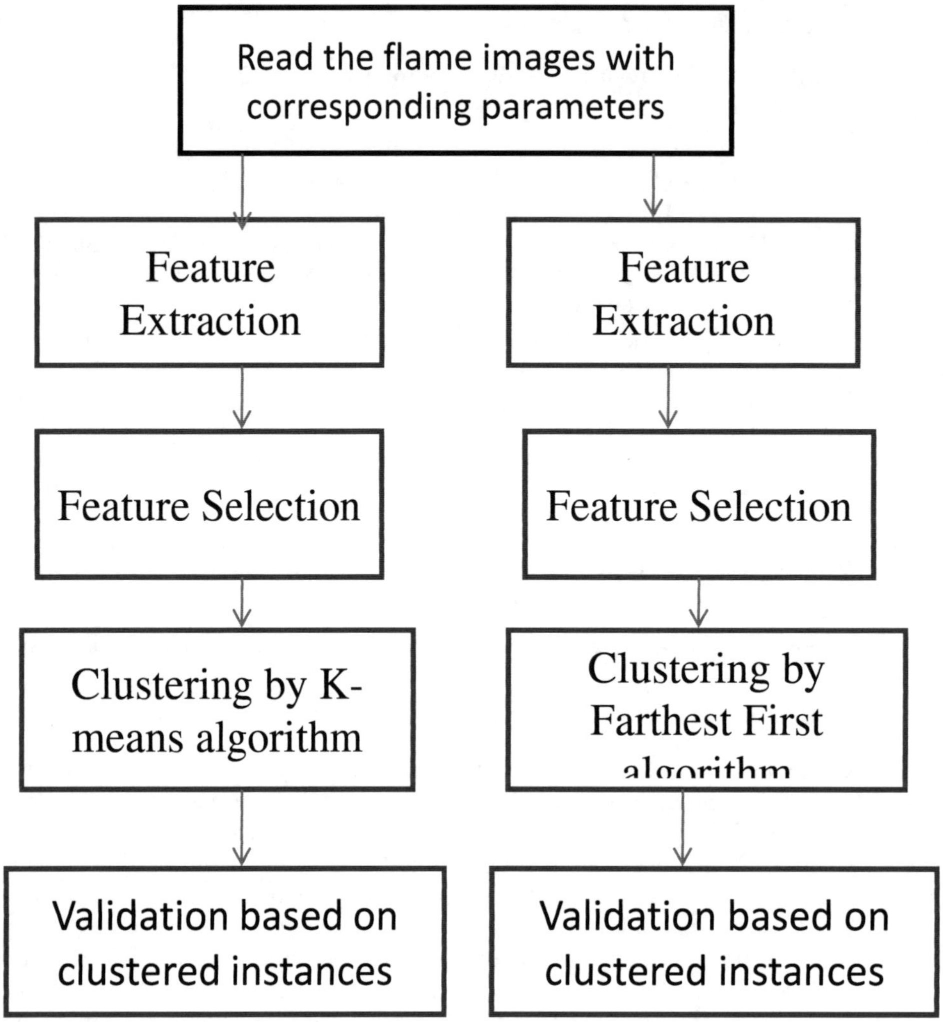

Fig.1 Block diagram for implementation of imaging technique

3. IMAGE PROCESSING SYSTEM FOR FLAME MONITORING

Figure1 shows the block diagram for flame image analysis. Noise removal, which is a pre-processing technique to remove noise from Flame images, is done using median filters. The attributes included maximum intensity, minimum intensity, area of the flame, standard deviation; Kurtosis and Internal Density were extracted using Image J. Table 2, shows the feature elimination using the mean threshold. Similarly, the Table 3 shows the second level of threshold in which the 1st feature was eliminated using the above said criteria. As a result out of nine features only seven features were selected and used for classification. The main aim of clustering was to identify the combustion condition [5, 6]. The classification was done with Weka, a user-friendly tool. The procedure for implementing any clustering algorithm to find the combustion quality is shown in Figure1.

Table.2 Features selected after first level of threshold

S. No	Mean	Features selected
Mean Threshold is 1999878		
1.	76576.9	1
2.	233.5838	2
3.	45.17203	3
4.	254.4443	4
5.	18.99749	5
6.	254.7211	6
7.	151.3636	7
8.	119.8586	8
9.	17921246	0

Table.3 Features selected after second level of threshold

S. No	Mean	Features selected
Mean Threshold is 9735.008		
1	76576.9	0
2.	233.5838	2
3.	45.17203	3
4.	254.4443	4
5.	18.99749	5
6.	254.7211	6
7.	151.3636	7
8.	119.8586	8

4. RESULTS AND DISCUSSION

It is evident from Figure 2 that the Load generation can be tracked from the colour changes taking place in the flame during combustion and the intensity ranges from 230 to 240 nearly. In addition to it Table 4 and Figure 3 indicate the tracking of oxygen content. It is indicated that if the air fuel ratio is maintained closer to 4:1 then the mean intensity of the flame is increased which indicates that the flame colour approaches to yellowish white indicating the complete combustions. From Table 4 and Figure 3 it is identified that, the load generated is uniform at higher intensities.

Table.4 Tabulation for mean intensity Vs % Average O_2

S.No	Average Intensity	% O_2 (Left)	% O_2 (right)	Supply of Oxygen (%)
1.	239.052	4	3.8	3.9
2.	235.6469	3.6	4	3.8
3.	234.751	3.4	3.9	3.65
4.	233.8803	3.2	3.4	3.3

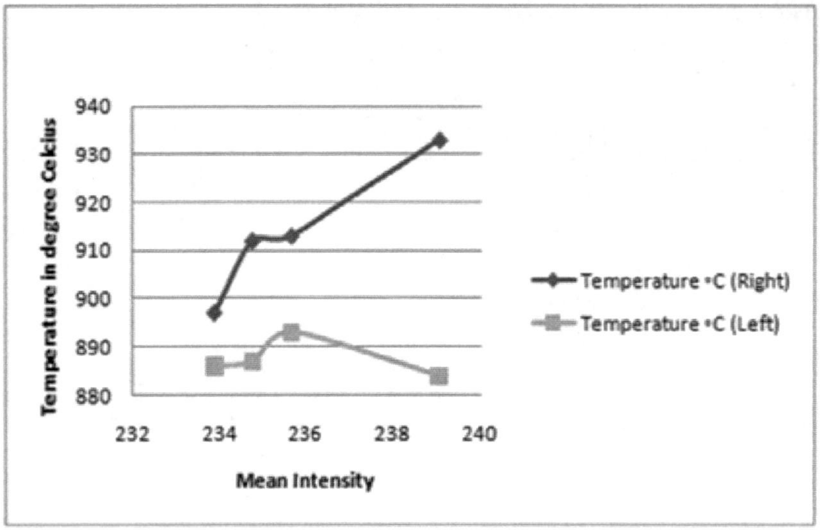

Fig. 2 **Scatter plot for % Air Supply**

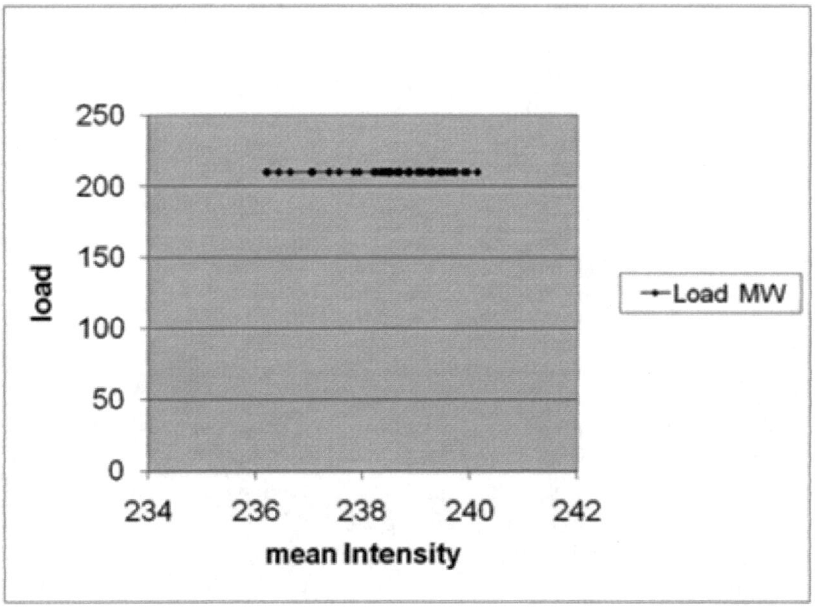

Fig.3.**Tracking of Intensity-load change**

Additionally, it is obvious that high temperature of the flame is related to colour mapping table, which is used to identify the temperature in various portions of the flame images [7]. The difficulty involved is that the colour associated with each portion of the flame need to be identified using segmentation algorithms with appropriate threshold [8]. The clustering results (Table 5 and 6) are represented graphically as shown in Figure 4. In Figure 4 shown above the farthest first algorithm applied for reduced feature set by the proposed technique also yields cent percent results to identify the combustion conditions.

Table 5. Comparison of various clustering algorithms

S.No	Parameters for comparison	K-means clustering with all features	K-means clustering with reduced feature set	Farthest First clustering with all features	Farthest First clustering with reduced feature set
1.	Number of Attributes	10	09	10	09
2.	Number of iterations	24	21	24	20
3.	Squared error in %	87.2% misclassified	86.3% misclassified	No misclassification	No misclassification

Table 6. Comparison of Performance of different clustering algorithms

S.No	Algorithm Type	% Performance classification
A1.	K-means with all features	13
A2.	K-means with reduced features	14
A3.	Farthest First with all features	100
A4.	Farthest First with reduced features	100

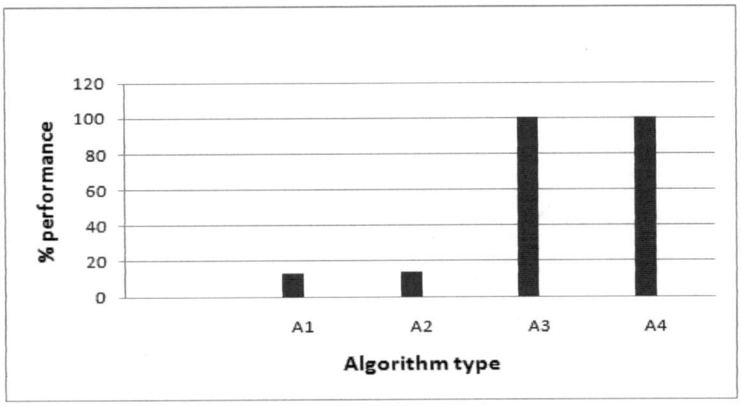

Fig. 4. Comparison Chart forClustering Performance

5. CONCLUSION

The clustering of images was done to identify the combustion quality. The performances of the various clustering algorithms were shown in Figure 4. Therefore the automation of combustion quality and power generation can be made possible by flame image analysis. The work can be extended to develop an intellectual system to monitor and control the air fuel ratio. It is also possible to reduce flue gas emissions such as to minimize the air pollution.

References

[1] Gang Lu a, Yong Yan a, Steve Cornwell b, Michael Whitehouse b, Gerry Riley b, *Impact of co-firing coal and biomass on flame characteristics and stability,* Elsevier, July 2007, pp: 1133–1140.

[2] Dr. G. Lu and Prof. Yan, Advanced Monitoring and Characterization of combustion flames, *20th annual meeting and meeting of the Advanced Power Generation Division*, University of Leeds, April 2009.
[3] C K Tan, S J.Wilcox, J Ward and M Lewitt*Monitoring near burner slag deposition with a hybrid neural network system*, MEASUREMENT SCIENCE AND TECHNOLOGY, April 2003.
[4] G Gilabert, G Lu and Y Yan,Three dimensional visualisation and reconstruction of the luminosity distribution of a flame using digital imaging Techniques, Sensors & their Applications, 2005, pp:167–171.

[5] M.G. Abdul Rahman, J.R. Gibbins, A.K. Forrest, Combustion in Power Station Boilers – *Advanced Monitoring Using Imaging,* First published 2004 © Imperial College of Science, Technology and Medicine copyright 2004.
[6] Pu Han, Xin Zhang, Chenggang Zhen and Bing Wang ,Boiler Flame Image Classification Based on *Hidden Markov Model, IEEE ISIE 2006*, July 9-12, 2006.
[7] Enrique Teruela, CristóbalCortésb, Luis Ignacio Díezb, InmaculadaArauzob, Monitoring andprediction of fouling in coal-firedutility boilers using neural networks, Chemical Engineering Science 60, 2005 5035 – 5048.
[8] Susanne Paulruda, CalleNilsson,The effects of particle characteristics on emissions from burning wood fuel powder , Fuel 83, 2004.
[9]. K. Sujatha, N.P.G. Bhavani, V. Srividhya, V. Karthikeyan, N. Jayachitra, Soft sensor with shape descriptors for flame quality, prediction based on LSTM regression, Real-Time Data Analytics for Large Scale Sensor Data. https://doi.org/10.1016/B978-0-12-818014-3.00006-1, 2020.

An Integrated Approach for Flood Prediction by Using Block Chain Network and Machine Learning

Kalpana Murugan [1], A.Seeta Reddy[2], M.Sivasainath Reddy[3], P.Raviteja Reddy[4]

[1] Department of Electronics and Communication Engineering, Kalasalingam Academy of Research and Education, Krishnan Koil, Virudhunagar-626126, Tamilnadu, India

[2,3,4] Department of Electronics and Communication Engineering, Kalasalingam Academy of Research and Education, Krishnan Koil, Virudhunagar-626126, Tamilnadu, India

E-mail: drmkalpanaece@gmail.com

Abstract. Now a days heavy rainfall becomes the major problem in India because of the changes that occur in global environment and in climatic conditions this problem rises so, To overcome this problem, this proposed method i.e advanced flood alerting system (AFAS) had been introduced by using machine learning and block chain network. This method helps to predict an early occurrence of flood with an efficient output probability of prediction. The output of this project gives an accurate and approximate probability percentage of flood prediction. After predicting the flood it gives an alert or caution to that particular area that has been effected by flood so, it helps them to take necessary precaution steps to follow ,If this happened then the victims that effected by flood can be saved and also death loss and property loss can be reduced in an effective manner.

1. Introduction

Now a day's floods becomes the major natural disaster so, Predicting the rainfall prior makes to stop the floods and saves the live of humans, animals and properties. In this [1] exact rainfall prediction is proactive task for engineers and scientists. Several methods are used for rainfall prediction for weather forecasting with circumstances and several neural network algorithms approaches used for heavy rainfall in advance approach of detecting the flood in the prediction by various engineer's scientists in current generation. In [2] output of the flood accuracy is very less because today it is challenging and complex task. Among all approaches Machine Learning is latest technology for prediction purposes so that using this technology to predict the rainfall earlier. In [3] many methods are available for prediction but accuracy is low, when coming to machine learning it works like human brain and output accuracy is more compare to other methods. In [4] when provide data to ML, give the training to it then it processes data and works according to given input instructions. Occurring many disasters damage the properties, human loss and decrease the country's economy, in those disasters flood disaster is great impact to the loss.

IOP Conf. Series: Materials Science and Engineering 1049 (2021) 012016 doi:10.1088/1757-899X/1049/1/012016

2. LITERATURE REVIEW

In [5] accuracy of rainfall forecasting is major problem in countries like India whose economy mainly depends on agriculture. Statistical techniques failed to predict the good accuracy but non linearity of rainfall data by artificial neural network makes good accuracy of rainfall prediction. Predicting the rainfall earlier stage help to avoid the floods and save the human lives. In [6] collect data in real time by using internet of things and it transfers data to computer device through big data and by using internet of things they have done the approach of flood prediction they have used sensor to overcome the issue. In [7] predicts the heavy rainfall by using the conversation of the soil and the basin techniques has been implemented so this project may not have high probability of prediction. In [8] proposed heterogeneous wireless sensor network to predict the floods in effective way. Hydraulic spectralsatellitsystem has been used in geographical information concept This cause the high range of complexity. In [9] this model they have developed the algorithm and this algorithm has been introduced in python programming language so, this concept of input pattern to the python has been very complicated like using in prediction model and forecasting model. In [10] proposed flood modelling and rainfall-runoff relationships has been presented by taking parameters like rainfall data, temperature, soil moisture and watershed. Rainfall runoff defines the amount of water falls in a certain period of time. Flood prediction is defining precipitation data to predict the water levels in timeline model. In [11] predicting the rainfall by using the digital cloud images clustering technique analysing the colour. Dimension of images are 400 x 250 taken from the internet and types of cloud estimated using the clustering methods and finally give the status of rainfall. In [12] discuss the machine learning techniques to predicting the rainfall in earlier. Heavy rainfall causes damage to crops yield due to this decrease the country economy to overcome this problem taking the previous rainfall according to crops seasons like rabbi, Kharif and predict the rainfall in future which crop would be suitable in occurrence of more rain. In [13],[14] solved the problems of short- term rainfall forecasting from the DRCF greedy algorithm using machine learning. External results show that overall performance of the art approaching which intervals on three hours. The greedy algorithm can only determine the optimal structure of local MLP, but could not global optimal structure. In [15],[16] to find another concept of an idea to predict the heavy rainfall network of geographical wavelet through satellite communication via gsm module the alert can be send through the mobile phone of weather department, this makes more complex and cost. The validation performance of the models is valued by the numerical methods. In [17] the artificial approach of deep learning techniques has been implemented in advance model but it is so complex and vast thing of neural based system. In [18] this model of flood prediction system the existing method has been some drawbacks of using basin proposed method, in order to overcome this proposed method has been introduced. In [19] this wirelesssensor network for flash flood warning they have used mat lab on neural network using nodes and sensors has been used to solve the problem. In [20] assuming floods as stochastic processes, they can be predicted using certain probability distributions from historical streamflow data.

3. PROPOSED METHOD

3.1. An Advanced Flood Alerting System (AFAS method)

Many people in India has been subjected to the effects of climatic conditions like natural calamities i.e. tsunami, heavy flood and heavy rainfall. Due to this all disasters that occur in India the people who were living in India has been suffering a lot so, in order to give solution to this type of problems, Came to an idea i.e. 'Early detection of flood forecasting system' This proposed method has been introduced to find solution for above issue mentioned. Our proposed method (AFAS) is that predict an early occurrence of flood and after prediction making an alert to the public that there may be an occurrence of flood in your city, by giving an alert or cautioning them, they can able to take safety and precaution measurements. The main motive of this project idea is that by using this idea and approach at least one mankind death loss can be reduced. In this machine learning and block chain technology has been used. Machine learning is an algorithm that has an ability to learn from the past experiences and learn itself it is an advanced model and method that are going to use in this project, this helps the data to process and compare the data that are taken i.e. previous year data it will compare the year by year by taking all ten years of rainfall data in June, July and august i.e. (past ten years data). After processing the data of ten years by taking the mathematical average values and the genetic decision making tree graph of all the possible methods to find an average value and decision making figure i.e. value or can be defined as probability percentage of mathematical value that is going to make a decision or coming to an conclusion to prediction of flood probability percentage value of the processed data. By using this machine learning techniques accuracy and efficiency of the output will be increase.Table.1. shows the previous year's data of weather parameters like temperature, humidity and pressure from the year 2010 to 2019 and only in the months of June, July and August. The rain fall data has been collected from the weather forecasting department. Table.2. Shows the year wise rainfall data obtained from an India Meteorological Department from 2010 to 2019.By AFAS method predicts the flood in the months of June, July, August of current year 2020. Finally, the prediction data has been forwarded to the weather forecasting department by using block chain network that is used for security purpose. Block chain network has been used here in this project for security purpose and after making the decision of flood prediction the data has to be secured in a very confidential manner in order to avoid the miscommunication of data. This act as a barrier of security line between the sender and receiver.

Table 1. Weather parameters

				TEMPERATURE (^{O}C)				HUMIDITY (%)				PRESSURE (mbar)			
S.NO	STATE	CITY	YEAR	JUNE	JULY	AUGUST	AVG	JUNE	JULY	AUGUST	AVG	JUNE	JULY	AUGUST	AVG
1.	KERALA	KOCHI	2010	27	26	26	26.3	88	90	88	88.6	1009	1009	1009	1009
2.	KERALA	KOCHI	2011	27	26	26	26.3	88	89	89	88.6	1009	1009	1009	1009
3.	KERALA	KOCHI	2012	27	27	27	27	86	85	86	85.6	1009	1009	1010	1009.3
4.	KERALA	KOCHI	2013	26	26	27	26.3	90	89	86	88.3	1009	1009	1010	1009.3
5.	KERALA	KOCHI	2014	28	27	26	27	86	87	88	87	1009	1010	1010	1009.6
6.	KERALA	KOCHI	2015	28	27	28	27.6	86	85	82	84.3	1009	1010	1010	1009.6
7.	KERALA	KOCHI	2016	27	27	28	27.3	87	85	83	85	1010	1010	1010	1010
8.	KERALA	KOCHI	2017	27	27	27	27	88	86	85	86.3	1010	1010	1009	1009.6
9.	KERALA	KOCHI	2018	27	26	26	26.3	88	89	87	88	1009	1010	1010	1009.6
10.	KERALA	KOCHI	2019	28	27	26	27	82	84	87	84.3	1008	1009	1010	1009
11.	**AVERAGE**			27.2	26.6	26.7	26.83	86.8	86.9	86.1	86.8	1009.1	1009.5	1009.7	1009.4

Table 2. Rainfall data

IOP Conf. Series: Materials Science and Engineering 1049 (2021) 012016 doi:10.1088/1757-899X/1049/1/012016

RAINFALL							
S.NO	STATE	CITY	YEAR	JUNE	JULY	AUGUST	ANNUAL
1.	KERALA	KOCHI	2010	667.5	629	356	1652.5
2.	KERALA	KOCHI	2011	788.5	536.8	492.7	1817.5
3.	KERALA	KOCHI	2012	430.3	362.6	501.6	1294.5
4.	KERALA	KOCHI	2013	1042.7	830.2	369.7	2241.9
5.	KERALA	KOCHI	2014	455.4	677.8	733.9	1867.1
6.	KERALA	KOCHI	2015	563.6	406	252.2	1221.8
7.	KERALA	KOCHI	2016	482.4	804	432.6	1719
8.	KERALA	KOCHI	2017	705.9	966	489	2160.9
9.	KERALA	KOCHI	2018	469	924.7	349	1742.7
10.	KERALA	KOCHI	2019	438.2	503	269.3	1210.5
AVERAGE				604.35	664.01	424.96	1692.81

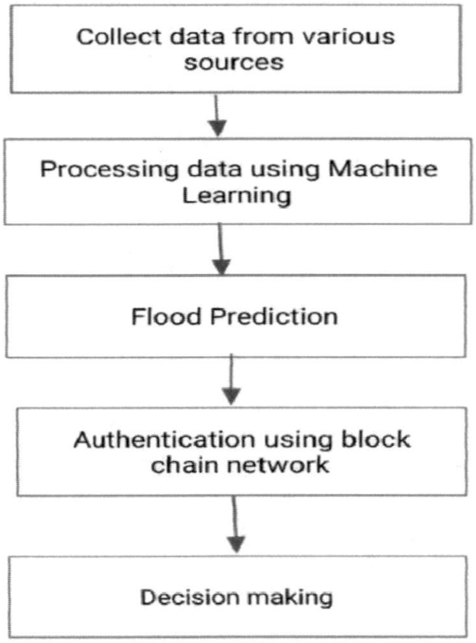

Figure 1. Flow Diagram of AFAS Method

Flow Diagram description:

Figure.1.depicts the flow diagram for AFAS method. First identifying the problem in a particular city or in an area and collect the previous ten years data in that city from the various source and this data has been processed by using machine learning algorithm and dataset. By the concept of machine learning techniques the processed data has been compared year by year and come to a conclusion that the prediction take place or not and if it happens how much amount of the probability will occur if the probability of prediction is high then it comes to a conclusion that there is an early occurrence of flood in that city. Finally, the prediction data has been forwarded to the weather forecasting department by using block chain network that is used for security purpose. Block chain network has been used here in this project for security purpose and after making the decision of flood prediction the data has to be secured in a very confidential manner in order to avoid the miscommunication of data. This act as a barrier of security line between the sender and receiver.

4. RESULTS AND DISCUSSION

Flood is one of the major problem that occurs due to natural calamities, An Advanced Flood Alert System (AFAS) is proposed to predict the heavy rainfall and protect the lives of humans, animals, and properties. In order to overcome the drawback this work proposes the AFAS method to predict the rainfall using the machine learning algorithm and block chain technology. AFAS method make an alert to the public and provide a caution in an earlier or before the occurrence of flood. So the people can take necessary safety measurements to overcome the flood. The result shows that the AFAS method has more accurate than the Artificial Neural Network (ANN) method in terms of rain fall prediction. By using this project an early occurrence of flood can be detected and this result i.e. the probability of prediction can be used to make an alert to the public in order to avoid the death and property loss. In Anaconda software by using Jupyter notebook the data set has been used which inserted in separate file to access this file

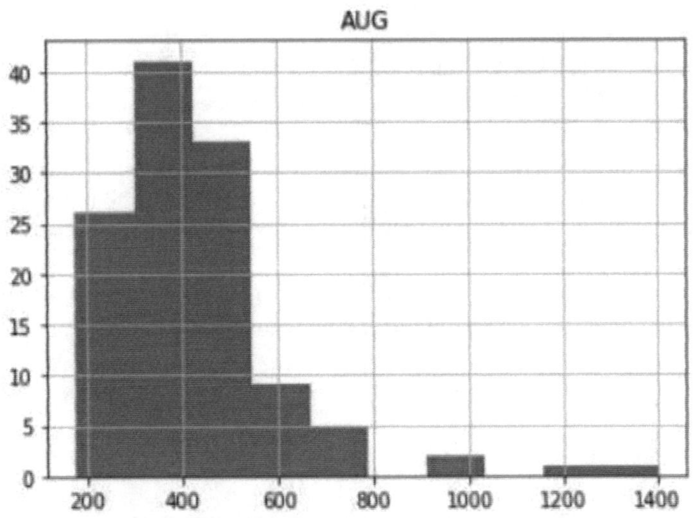

Figure 2. Rainfall Prediction graph of June month

Import function be used then using our method that has been introduced in this project predicts the flood in 2020.Observe that June, July months prediction graph as show in fig 2,3,4 with the normal rainfall values but in August month prediction graph shows higher rainfall so there is chances to occur the flood in the month of august in Kerala. Figure.5. shows the performance level graph for the AFAS and ANN method.

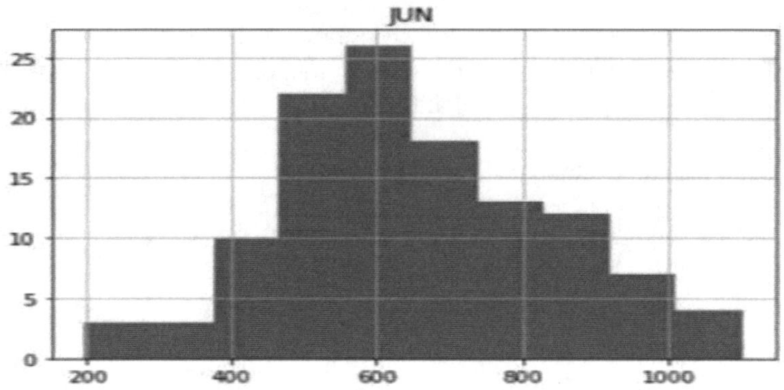

Figure 3. Rainfall prediction graph of July month

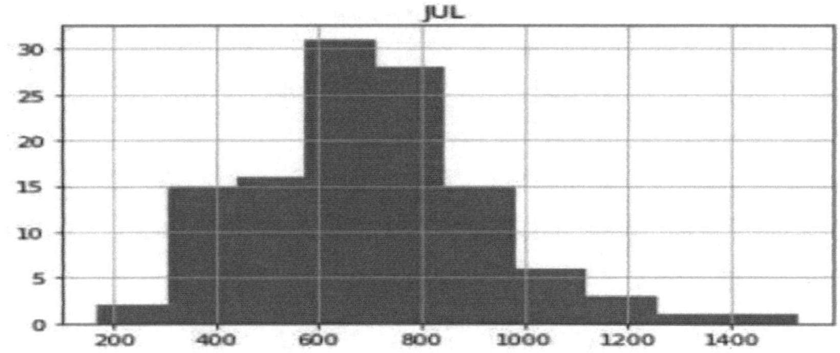

Figure 4. Rainfall prediction graph of August month.

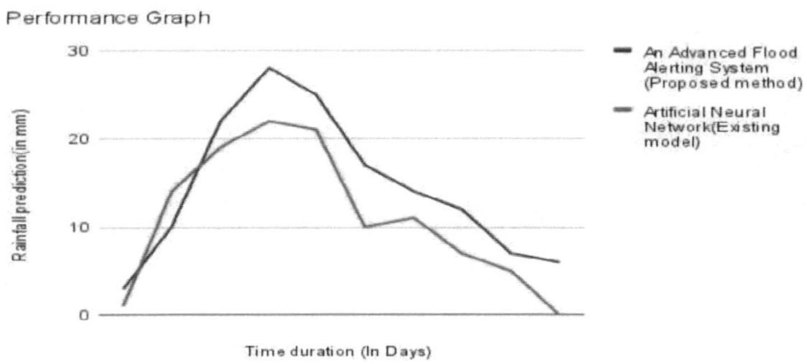

Figure 5. Performance of Rainfall prediction in AFAS and ANN

5. CONCLUSION AND FUTURE WORK

This proposed method (AFAS) predicts the rainfall prior makes to stop the floods and saves the live of humans, animals and properties. Exact rainfall prediction is proactive task for engineers and scientists. Several methods are used for rainfall prediction for weather forecasting with circumstances and several neural network algorithms approaches used for rainfall prediction by various engineers and scientists in current generation. In order to save the people from natural disaster like floods and heavy rain fall the prediction concepts has been introduced and this (AFAS) method predicts an early occurrence of flood has been useful for the areas where flood occurrence is more and in order to solve the problem of heavy flood that causes the human death and property loss, this project helps to make an alert in the public and make a caution in early or before occurrence of flood so, that the people can take necessary safety measurements to overcome the flood.

References

[1] Akhtar, N., Rehman, A., Hussnain, M., Rohail, S., Missen, M. S., Nasir, M., ... & Pasha, M. (2019). Hierarchical coloured petri-net based multi-Agent system for flood monitoring, prediction, and rescue (fmpr). *IEEE Access*, *7*, 180544-180557.

[2] Fleming, S. W., & Goodbody, A. G. (2019). A Machine Learning Metasystem for Robust Probabilistic Nonlinear Regression-Based Forecasting of Seasonal Water Availability in the US West. *IEEE Access*, *7*, 119943-119964.

[3] Puttinaovarat, S., & Horkaew, P. (2020). Flood Forecasting System Based on Integrated Big and Crowdsource Data by Using Machine Learning Techniques. *IEEE Access*, *8*, 5885-5905.

[4] Parmar, A., Mistree, K., & Sompura, M. (2017, March). Machine learning techniques for rainfall prediction: A Review. In *International Conference on Innovations in information Embedded and Communication Systems*.

[5] Venkatesan, C., Raskar, S. D., Tambe, S. S., Kulkarni, B. D., & Keshavamurty, R. N. (1997). Prediction of all India summer monsoon rainfall using error-back-propagation neural networks. *Meteorology and Atmospheric Physics*, *62*(3-4), 225-240.

[6] Sahai, A. K., Soman, M. K., & Satyan, V. (2000). All India summer monsoon rainfall prediction using an artificial neural network. *Climate dynamics*, *16*(4), 291-302.

[7] Philip, N. S., & Joseph, K. B. (2001, May). On the predictability of rainfall in Kerala-An application of ABF neural network. In *International Conference on Computational Science* (pp. 400-408). Springer, Berlin, Heidelberg.

[8] Philip, N. S., & Joseph, K. B. (2003). A neural network tool for analyzing trends in rainfall. *Computers & Geosciences*, *29*(2), 215-223.

[9] Akyildiz, Ian F., Weilian Su, Yogesh Sankarasubramaniam, and Erdal Cayirci. "A survey on sensor networks." *IEEE Communications magazine* 40, no. 8 (2002): 102-114.

[10] Seal, V., Raha, A., Maity, S., Mitra, S. K., Mukherjee, A., & Naskar, M. K. (2012). A simple flood forecasting scheme using wireless sensor networks. *arXiv preprint arXiv:1203.2511*.

[11] Sehgal, V., & Chatterjee, C. (2014). Auto updating wavelet based MLR models for monsoonal river discharge forecasting. *Int. J. Civ. Eng. Res*, *5*, 401-406.

[12] Bakhtyari-kia, M., Pirasteh, S., Pradhan, B., Mahmud, A. R., Sulaiman, W. N. A., & Moradi, A. (2012). An Artificial Neural Network Model for Flood Simulation Using GIS. *Johor River Basin, Malaysia. Environmental Earth Sciences*, *67*(1), 251-264.

[13] Fang, S., Xu, L., Pei, H., Liu, Y., Liu, Z., Zhu, Y., ... & Zhang, H. (2014). An integrated approach to snowmelt flood forecasting in water resource management. *IEEE transactions on industrial informatics*, *10*(1), 548-558.

[14] Tuxworth, G. (2016). Automated Morphology-based Analysis of Neural Cells.

[15] Abhishek, K., Kumar, A., Ranjan, R., & Kumar, S. (2012, July). A rainfall prediction model using artificial neural network. In *2012 IEEE Control and System Graduate Research Colloquium* (pp. 82-87). IEEE.
[16] Mishra, N., & Jain, A. (2014). Time Series Data Analysis for Forecasting–A Literature Review. *International Journal of Modern Engineering Research*, *4*(7), 1-5.
[17] Lee, S., Cho, S., & Wong, P. M. (1998). Rainfall prediction using artificial neural networks. *journal of geographic information and Decision Analysis*, *2*(2), 233-242.
[18] Mitra, P., Ray, R., Chatterjee, R., Basu, R., Saha, P., Raha, S., ... & Saha, S. (2016, October). Flood forecasting using Internet of things and artificial neural networks. In *2016 IEEE 7th Annual Information Technology, Electronics and Mobile Communication Conference (IEMCON)* (pp. 1-5). Ieee.
[19] Castillo-Effer, M., Quintela, D. H., Moreno, W., Jordan, R., & Westhoff, W. (2004, November). Wireless sensor networks for flash-flood alerting. In *Proceedings of the Fifth IEEE International Caracas Conference on Devices, Circuits and Systems, 2004.* (Vol. 1, pp. 142-146). IEEE.
[20] Basha, E. A., Ravela, S., & Rus, D. (2008, November). Model-based monitoring for early warning flood detection. In *Proceedings of the 6th ACM conference on Embedded network sensor systems* (pp. 295-308).

IOP Conf. Series: Materials Science and Engineering 1049 (2021) 012017 doi:10.1088/1757-899X/1049/1/012017

Artificial Intelligence (AI) adaptation: Analysis of determinants among Small to Medium-sized Enterprises (SME's).

Sanjeev S Ingalagi[1], R R Mutkekar[2] and P M Kulkarni[3]

[1] Corresponding author
[1,3] KLS, Gogte Institute of Technology, Department of MBA, Belagavi, Karnataka, India
[2] Goa Institute of Management, Sattari, Goa, India

E-mail: ssingalagi@git.edu, rohitrm@gim.ac.in, pmkulkarni@git.edu

Abstract. Implementation of Artificial Intelligences changes the way of doing business in small and medium enterprises. The determinants for success or failures for implementation of artificial intelligences systems in small firms remain very unclear in India. Hence, the paper attempts to analyse various perceptive of AI implementations from a management point of view. Data analysis was carried out of 167 small and medium firms located in Karnataka state (India) from two tier towns .Existing literature review revealed top management's commitment, organisational readiness, external support, employee adoption and competitive pressure influence adoption of artificial intelligence technologies substantially. The study confirms influences of AI technology in small and medium firm's management. Clear understanding of drivers for AI implementation helps managers and owners of SMEs to take up appropriate initiatives. The results provide assorted perceptive to AI services providers to integrate business proposals for small and medium enterprises needs.

1. Introduction:

Artificial intelligence (AI) has changed the way businesses operate. AI includes techniques that can analyse patterns in raw data for making correct business decisions. Business decisions in SME's are related to production, marketing, supply chain, human resource and finance. Generally, AI analyses employees data, customer interactions, production and financial data interactions stored within the organisation's system. Analyses report provides actionable results for managers.AI analyses both structured and unstructured data sets using algorithms.

Past studies acknowledge AI technologies improvise organisational efficiency and performance.AI technology delivers quality business value to customers. Researchers in past [1] contend that "large organisations use integrated AI systems for advanced purposes such as predicting future sales trends in the marketing, evaluating customer's preferences, choosing profitable business activities for making financial investment based on returns".

Although, small and medium scale enterprises form basic economic structure of the country, they lag far behind in utilising AI. The lack of basic understanding of AI benefits and limited resources for AI adaptation are main reasons for lack of AI integration into SME's. The number of studies related to AI and SME's are scarce. Hence, the study draws on influences of external and internal factors on AI implementation in SME's. Extensive research on the impact of AI adaptation has been done in past .However generalised findings cannot be applied to SME's as they operate in different ecosystems.

2. Literature Review:

Artificial intelligence referred as machine intelligence demonstrated by machines unlike intelligence displayed by human beings. The term AI is used to illustrate machines that impersonate cognitive capability. Machines can intelligently communicate with each other by passing on signals. Signals are generated by machines through machine learning. A machine learns to processes data to make a decision.AI can be as "an umbrella term consisting of cyber –physical systems, big data analytics, Automation and machine learning. The AI of today is heavily reliant on collection, usage and processing of big data [2]".

Customer's demands keep changing over the period of time. Hence, firms need to keep up with changing customer demands. In order to be relevant to customer's demands, firms have to transform themselves as agile and cost efficient by reducing operational cost.This can be achieved by digitalisation in supply chain. There has been increasing integration of marketing, finance, human resource and production departments. Inter departmental transactions produce huge data with different volume, variety and velocity. The transactional data is saved as structured and unstructured data sets. AI involves integration of physical and virtual systems acting as a whole. The systems are connected through intelligent sensors communicating seamlessly with each other to meet a pre defined goal. AI has a potential to change way of doing business .Hence, implementation of AI in a firm is the major strategic decision.

AI maximises operational efficiency and transparency .Hence, its strategic implementation provides stimulus to firms. Firms implementing AI have to convert data into intelligent and understandable versions so that machines can make sense out of data. AI systems enhance agility of the organisations by speeding up data processing and decision making.AI has significant influences on different kinds of industries. The majority of large manufactures use AI to manufacture advanced equipments.

Table 1: Literature Summery.

Sr.No	Author and Year	Details	Outcomes
1	(Chen et al.,2015;Lai et al., 2018); (Priyadarshinee et al., 2017)	Top management commitment to the implementation of AI is necessary as they are the decision makers for creating technological ecosystem.	Top management commitment was considered as as an independent variables to study AI adoption in the study
2	(Chen et al., 2015) (Gangwar, 2018)	Organisation Readiness measures the extent of IT infrastructure available in the firm . Additionally it also includes skilled man power availability and required finances	Organisation readiness was considered an an independent variable to measure AI adoption in the model.
3	(Ghobakhloo,M.,Arias Aranda,D., & Benitez-Amado, J. (2011).	External support includes support from third party vendors who specialise in implementation of AI technology is necessary to speed up the process of AI implementation as SME's lack skilled human resource	External support has been considered as an independent variable influencing AI implementation
4	(Lai et al., 2018)	It is important for the SME's to survive in the business as large manufacturers choose SME's using advanced manufacturing technology to outsource production orders.	Competition pressure forces SME's to go for technical up gradation
5	(Lai et al., 2018; Xu et al., 2017)	Employees of SME need to be skilled in order to implement and maintain AI	Complexity of adoption of AI makes it less attractive and employee skillset helps in

		technology	moderating AI adoption.
6	(Ren, Wamba, Akter, Dubey, & Childe, 2017; Wamba et al., 2017)	AI based technologies gives a stimulus to improvise firms efficiency and effectiveness due to high strategic benefits and operational benefits	AI adoption for a SME's benefits in terms of operational efficiency
7	(Wamba et al., 2017).	AI applications are not just limited to manufacturing firms. AI can be adopted for services based industries as well. AI is a very versatile technology for efficiency improvement	AI can be adopted for manufacturing and service based firms.
8	(Alharbi et al. 2016)	The researchers mention that variation in the country requirements and other environmental factors determine AI adoption.	AI implementation is dependent upon country specific conditions and environmental conditions hence AI implementation changes as per requirements of different regions.
9	Ghasemaghaei (2018)	The author argues that AI implementation doesn't guarantee improvised performance	AI implementation not just an end in itself, more than that utilisation of AI in right kind of process is significant. AI utilisation should happen where firms can generate maximum business value.
10	(Buonanno et al., 2005)	There is a significant difference among in firms in terms of organisational resources , organisational structure and technological infrastructure .	There is variance in terms of organisational resources , organisational structure and technological infrastructure even within SME's

3. Hypothesis formulations:

Past studies assert that "top management's involvement being crucial for implementation of new technologies in the SME's.The continuous support of top management is necessary for organisation readiness [14]". Implementation of new technologies requires utilisation of organisation resources .Top management / owners of SME's have ability to divert organisational resources as they are decision makers. Top management vision determines level of AI adaptation and entire organisation ecosystem. Following hypothesis can be deduced from above conceptualisation.

H1: SME's Top management commitment significantly influences artificial intelligent adaptation.

"Organization readiness indicates preparedness of firms to absorb AI systems into its processes" as per mentioned in literature [6]. Organization readiness includes firm's IT infrastructure, financial and skilled human resource to implement new technology. "As per previous studies done in academics, organizational readiness is a precondition for AI adaptation [6][13]".For SME's, organizational readiness has a positive influence on implementation of AI system. AI implementation will be faster in firms of high organisational readiness. The study proposes a hypothesis saying that organizational readiness is one of the critical preconditions for AI adoption.

H2: SME's have Organisational Readiness (OR) for AI implementation

External environment includes competitive pressure and external support. "Competitive pressure refers to influences of external environment such as rival small units, supply chain partners, large manufacturers on small and medium scale units to adopt AI[1]".It is a pressure to comply with customers, suppliers and competitors. Many firms are expected to adopt new technology due to competitive pressure. Studies [14] reckon "competitive pressure significantly influences technology adoption in SME's. Pressure from media, competitors and customers significantly affect sustainable manufacturing practises". Some scholars believe increased usage of AI tools in the workshop enhances firm's ability to improve manufacturing efficiency and survivability in the competition. Hence, authors develop hypothesis as below.
H3: Competitive Pressure (CP) pushes SMES significantly for AI adaptation.

The support from external partners is very crucial for small and medium firms to improvise capabilities [1].Further, researchers reckon external support as "business environment entities such as business vendors, supply chain partner or information technology service providers to innovate existing processes within firms". The SME's needs to learn constantly from expertise of external service providers through training and open source platforms[6].As such, for many small firms lack workforce skill sets and lack IT infrastructure. It's crucial for SME's to get support from external services. Based on this perspective, researchers developed hypothesis as below.
H4: External Support (ES) influences significantly on AI adaptation

AI adaptation creates a novel working environment. Traditionally executed by workmen shall get migrated to AI systems. Hence, traditional skills will get redundant. Change in work environment will trigger workforce to acquire new skills of high order [15]. Employees who acquire new skills of higher order shall survive on jobs. Authors [16] explain employee adaptability as "ability of workers to get aligned to new age skills like IT and critical thinking". The areas for new skill requirements will be IT infrastructure, automation technology, data analytics and data security etc. Additionally, the dimension of adaptability changes in corresponding to firm's innovation cycle. The study focuses on perception of top management on AI adaptabilities of their employees. Top management ability to go for AI adaptation depends upon trust in their employee's ability to adopt AI. The mutual dependences between employee adaptability and successful AI integration lead to following hypothesis.
H5: Employee adaptability significantly influences implementation of AI technologies

4. Methods:
Structural equation modelling method was used for analysis of hypothesis .Exploratory factor analysis was performed followed by confirmatory factor analysis. Path model was used for analysis of hypothesis using regression coefficients.

4.1 Measurements:
The research method provides roadmap for research work to be carried out. The study utilises five point likert scale with options ranging from 1(strongly disagree) to 5(strongly agree). All six variables of model are measured this way. The draft version of measurement scale has been pretested by two academicians and four small industries owners. Based on their input, the questionnaire was revised for usage of pilot study to check understandibility. The Cronbach's alpha for all variables was more than 0.7 showing relatively good construct reliability (Table 3).

Table 2: Respondents Profile.

Sr	Number of Employees	N	Percentage
1	5–10 employees	26	15.56

2	11–49 employees	99	59.28
3	50 to 99 employees	34	20.35
4	100–149 employees	8	0.04
	Total	167	100%

The proposed model was conceptualised based on six influential variables classified as internal factors (top management commitment, organisational readiness, and employee adaptability) and external factors (competitive pressure and external support).

4.2 Sampling and data collection:
The respondents involved owners of small and medium scale enterprises in varied sectors such as machinery parts, Fabric metals, food and beverage, chemical, electrical machinery and office equipments (Table 2). SME's that have adopted artificial intelligence are the population for data collection. The sampling frame consisted of small and medium industries (total of 467)listed in Belgaum and Hubli by government operated district industries centre (DIC).The owners/managing directors of firms were targeted for the survey of questionnaire. The owners/managing directors of firms are the decision makers for AI adoption. It was presumed that owner's possess complete information of AI adaptation. The survey responses were collected through email and phone calls. The researchers used Common Method Bias (CMB) using Harman's single factor test to determine validity and reliability of the construct model. The first factor explained 37% of total variance. As the single factor does not carry the majority of explained variance, CMB was found to be not a serious issue in this case [17].

5. Data Analysis:
The proposed hypotheses were tested using structural equation modelling using standardised estimates. The method is combination of factor analysis and multiple regressions. Exploratory factor analysis (EFA) was carried out to be followed by confirmatory factor analysis(CFA) to arrive at measurement scales. The validity and reliability were tested for measurement model followed by structural modelling.

6. Results:
Post EFA and CFA, Authors established measurement model fit. The structural path model was established from measurement model. Structural model fit was performed to ensure accuracy of analysis.

6.1 Validity of measurement model:
To assess convergence of constructs, factor loadings, composite reliability (CR) and average variance explained were calculated (Table 4). Researchers [18] suggest values of CR, AVE and MaxR (Maximal Reliability) to be more than 0.7, 0.5 and 0.9 respectively". All constructs met threshold requirement and demonstrated acceptable construct validity. Discriminant validity was established by assessing (a) MSV < AVE, (b) Square root of AVE to be greater than inter-construct correlations (Refer Table: 3) .The table shows good discriminant validity.

*6.2 Assessment of structural model***:**
As per the table (5), the structural model showed good model fit revealing CMIN/DF = 1.585, which is below the cut off value 3.0. To identify a correct model based on collected data, multiple measures need to be used simultaneously [19]. Hence, multiple goodness of fit indices were calculated. It was found that GFI =0.926, CFI = 0.980 were extremely close to minimum threshold value 0.9 [18]. RMSEA = 0.039 was below threshold value of 0.05 which is good [20].As per researches[21], values between 0.08 to 0.10 can be considered as mediocre fit. RMSEA value below 0.05 is considered

goodness of fit. SRMR= 0.0381 is much near to threshold value of 0.05[20] establishing goodness of fit. The R square value for AI adopt was 0.59 showing a satisfactory model fit.

The researcher applied non parametric boot strapping with 4000 replications to assess structural model .As per Table (6), competitive pressure (β= 0.106; p< 0.05), Employee adaptability (β= 0.394; p<0.05), external support (β= 0.139; p<0.05) were significantly influencing AI adaptation. Top management support (β=0.001; p>0.05) and organisation readiness (β= 0.058; p> 0.05) were not significantly influencing AI adaptation.

Table 3: Factor Structure and Reliability of EFA.

Construct	Items	Cronbach's Alpha	Variance %	M	SD	Range
Competitive Pressure(CP)	4	0.938	37.073	3.98	0.701	0.784-0.987
Top Mgmt Commit(TMC)	4	0.924	13.122	2.25	0.91	0.838-0.916
Employee Adaptability(EA)	4	0.898	10.514	3.75	0.865	0.696-0.964
External Support (ES)	4	0.911	7.098	3.72	0.933	0.820-0.879
Organisation Readiness (OR)	4	0.9	6.307	2.99	0.947	0.662-0.938
Artificial Intelligence Adopt(AI)	4	0.846	4.283	3.92	0.708	0.614-0.781

Table 4: Validity of Measurement Model

	CR	AVE	MSV	MaxR(H)	ES	EA	OR	CP	TMC	AI
ES	0.913	0.723	0.263	0.914	0.850					
EA	0.892	0.676	0.496	0.956	0.351	0.822				
OR	0.894	0.680	0.246	0.971	0.496	0.241	0.825			
CP	0.940	0.796	0.304	0.982	0.463	0.532	0.327	0.892		
TMC	0.927	0.761	0.120	0.985	0.216	0.224	0.347	0.233	0.872	
AI	0.849	0.584	0.496	0.986	0.513	0.704	0.378	0.551	0.233	0.764

Table 5: Model Fit Summary

	CMIN/df	CFI	GFI	AGFI	SRMR	RMSEA	PCLOSE
Criteria	< 3	>0.95	>0.95	>0.80	<0.09	<0.05	>0.05
Model Indices	1.585	0.980	0.926	0.905	0.0381	.039	.992

Table 6: Model Summery (Regression Weights)

Hypotheses	Relationships	Estimate	S.E	C.R.	P	Results
H1	CP > AI	0.106	0.042	2.546	0.011	Significant
H2	TMC > AI	0.001	0.031	0.02	0.984	Insignificant
H3	EA>AI	-0.394	0.043	9.228	***	Significant
H4	ES > AI	0.139	0.038	3.689	***	Significant
H5	OR >AI	0.058	0.03	1.92	0.056	Insignificant

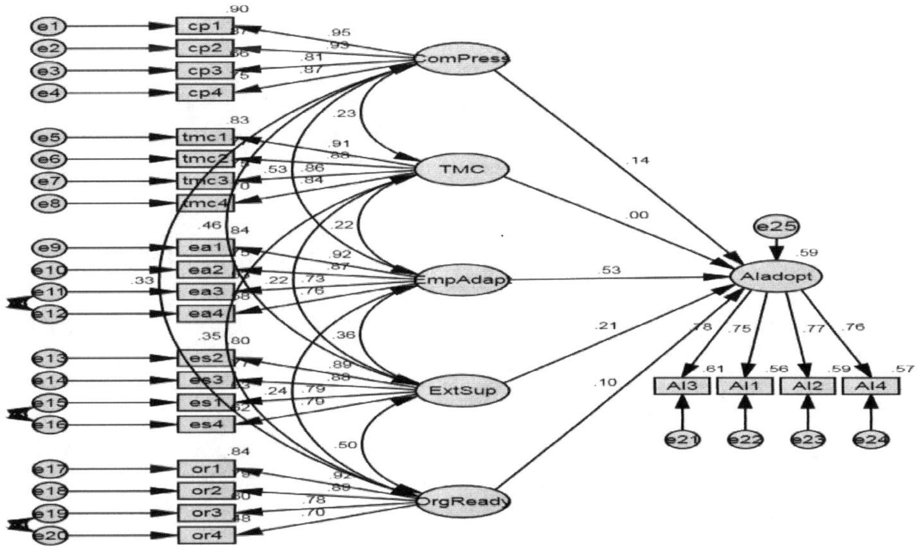

Figure: 1: Model Structure of Determinants Influencing on AI adaptation.

7. Discussion and Implications:

The influence of top management commitment on AI adoption is not supported by small and medium enterprises. The insignificant effect of top management on AI adaptation is inconsistent with findings in large firms as per [6][7]. However, insignificant effect is in line with findings of small and medium firms as per[1]. Top management (SME's owners) commitment is among the critical elements of technology adaptation. SME's owners are decision makers. Owner's vision determines AI adaptation. Top management support is decisive in creating new technology based ecosystem SME .The study finds SME's owners not being very supportive of AI adaptation due to lack of funds for investments. Additionally, authors noted a lack of government subsidies to speed up AI adaption in SME's industrial sector. However, SME's owners are optimistic of AI adaptation in years to come. The study finds non availability of AI expertise service providers in the market being another reason for slow AI technology absorption by SME's. The study notes that majority of top management of SME's being palpably unaware of AI benefits.

The study recommends AI expert service providers to illustrate successful case studies on AI implementation to top managements. SME's are unlikely to do risk investments in AI adaptation due to lack of visible benefits and data security being compromised [4].Therefore, communicating AI implementation benefits to top management could change their decision. AI services providers can enhance AI adaptation by providing services at a penetration pricing .Additionally; govt needs to provide incentives for SME's to build AI capabilities. Government can offer subsidised loan for building AI capabilities for SME's.

The substantial linkage is seen between organisational readiness and AI adaptation as previously noted by [6]&[7]. Past studies [5] emphasise that "It is not easy for SME's to adopt AI without adequate technological, financial and skilled human resources. A firm is unlikely to adopt AI if it doesn't have resources and capabilities". It is important for SME's to employee AI skilled employees for initial implementation and maintainace. To begin with technological, financial resources should be available to outsource AI implementation. Taking into consideration of paucity in financial resources, technology infrastructure, analytics capabilities and skilled resources can restrict AI adaptation in

SME's. AI service providers and governments need to play an active role in extenuating barriers. As a part of initial solutions, AI services providers need to train SME's employees on smooth initial integration to overcome skilled manpower shortage .Further, AI service providers can offer affordable services to SME's .For instances, providing a "pricing plan based on number of users enables SME's to invest an affordable amount by purchasing a minimum number of licenses required" as per [8]. AI services providers should assure SME's of technical support availability to tackle and enhance SME's capability.

Competitive pressure has a substantial linkage with AI adaptation. The insignificant effect isn't in congruence with findings mentioned in the past [5]. Most of SME's in Belgaum and Hubli are vendors for large manufacturers in bigger city like Pune and Bangalore. Large manufacturers give job work to SME's to reduce cost of manufacturing. A lack of multinational SME's in Belgaum and Hubli could be the potential reason for insignificance. Low level of AI services providers among local players could also be a potential reason for insignificance. Due to moderate infrastructure availability in Hubli and Dharwad, multinational companies have not invested in these towns. Hence, SME'S in twin towns seldom get affected by competitive pressure of globalisation. Low level of AI service penetration among local industries reduces competition. The large manufacturers prefer SME's who have good investments in AI for offering job work orders. The presence of multinational competitors in bigger cities of India and import of parts from other countries have played significant role in AI adaptation in local industries. As a matter of holistic development, Government needs to push development of industrial corridors in small towns .So far, government has been concentrating on developing infrastructure in bigger metro cities overlooking small town development.

As per research studies in [6], external environmental support speeds up AI adoption . The study is in line with findings of [6] enunciating a significant linkage between external support and AI adoption .If an AI service provider can reduce operational cost of a SME's by implementing AI solutions, SME's could most likely adopt AI. Lack of internal expertise in SME's could make AI services providers inevitable. Technical support offered by AI service vendors can reinstate lack of technical expertise for AI implementation as mentioned in [12]. Researchers in the past [1] reckon that "AI service providers need to provide trial versions to SME's and training to SME's employees to build their reputation". AI adoption is a high cost. Lack of human expertise makes AI implementation an impractical option for SME's. Information security is a barrier for SME's to hire an AI service provider. As such, social media engagement and referral system among SME's builds trust and brand for AI service providers.

The linkage between employee adaptability and AI implementation is significant. Implementation of Artificial intelligence creates a new working environment. Most of the work is done by artificial intelligent systems. Hence, traditional work values of SME's will be redundant. This will require employees to adopt newer skills sets as per [11]. The employees who adopt to new job requirements will survive. The researcher found significant and negative relationship between employee adaption with AI. The survey questions were asked to owners. Most owners of SME's presume that employees don't have adequate expertise to handle AI which is in consistent with findings in [6]. The study finds employee adaptability negatively and significantly influences AI adaptation. AI adaption is not easy for SME's as owners of SME's opine that existing human resource not having internal expertise. Should this be a case, outsourcing AI implementation would be a good option. to reduce cost, improve manufacturing efficiency and to attract cutting edge work orders from large manufacturers. SME's owners should consider AI technologies as strategic investment rather than expenditure because AI reduces operational costs, machine idle time. AI improves manufacturing efficiency helping small firms to attract new outsourced production work orders from large companies.

8. Future Directions:

Future studies can test this model for the larger firms. The study focused on small and medium industries located in two tier cities of southern Indian Karnataka state. Further, conceptual framework of the research required to be tested in other developing and developed countries. The results are affected due to lack of observation of dynamic changes in AI adoption. Further research can be based on a longitudinal study on established relationships mentioned in this study for an extended period of time. Other potential factors such as organisational culture, market pressure and technical infrastructure can be included in further studies.

9. Conclusion:

AI adoption improves effectiveness of the firms by using artificial tools to make intelligent managerial decisions. SME's are very essential for the growth of economy .However; SME's lack newer technologies. Considering the lack of study on the factor's driving the outcome of artificial intelligence on SME's in tier two towns, impact of organisational studies related to employee adaptability, top management commitment and organisational readiness were included in study. Impact of external factors such as competitive pressure, external support on AI adaption was investigated. Researchers reckon that external factors influence AI adaptation while SME's aren't ready internally to adopt AI. In future, govt interventions and affordability of AI services providers can play a significant role in dynamics of SMEs in AI adaption.

References:

[1]. Maroufkhani P, Tseng ML, Iranmanesh M, Ismail WK, Khalid H. Big data analytics adoption: Determinants and performances among small to medium-sized enterprises. International Journal of Information Management. 2020 Oct 1;54:102190.

[2]. Marr B. Artificial intelligence in practice: how 50 successful companies used AI and machine learning to solve problems. John Wiley & Sons; 2019 Apr 15.

[3]. Chen DQ, Preston DS, Swink M. How the use of big data analytics affects value creation in supply chain management. Journal of Management Information Systems. 2015 Oct 2;32(4):4-39.

[4]. Priyadarshinee P, Raut RD, Jha MK, Gardas BB. Understanding and predicting the determinants of cloud computing adoption: A two staged hybrid SEM-Neural networks approach. Computers in Human Behavior. 2017 Nov 1;76:341-62.

[5]. Ghobakhloo M, Arias-Aranda D, Benitez-Amado J. Adoption of e-commerce applications in SMEs. Industrial Management & Data Systems. 2011 Aug 30.

[6]. Gangwar H. Understanding the determinants of big data adoption in India: an analysis of the manufacturing and services sectors. Information Resources Management Journal (IRMJ). 2018 Oct 1;31(4):1-22.

[7]. Lai Y, Sun H, Ren J. Understanding the determinants of big data analytics (BDA) adoption in logistics and supply chain management. The International Journal of Logistics Management. 2018 May 14.

[8]. Xu W, Ou P, Fan W. Antecedents of ERP assimilation and its impact on ERP value: A TOE-based model and empirical test. Information systems frontiers. 2017 Feb 1;19(1):13-30.

[9]. Wamba SF, Gunasekaran A, Akter S, Ren SJ, Dubey R, Childe SJ. Big data analytics and firm performance: Effects of dynamic capabilities. Journal of Business Research. 2017 Jan 1;70:356-65.

[10]. Alharbi F, Atkins A, Stanier C. Understanding the determinants of Cloud Computing adoption in Saudi healthcare organisations. Complex & Intelligent Systems. 2016 Oct 1;2(3):155-71.

[11]. Ghasemaghaei M. The role of positive and negative valence factors on the impact of bigness of data on big data analytics usage. International Journal of Information Management. 2020 Feb 1;50:395-404.

[12]. Shamim S, Zeng J, Shariq SM, Khan Z. Role of big data management in enhancing big data decision-making capability and quality among Chinese firms: A dynamic capabilities view. Information & Management. 2019 Sep 1;56(6):103135.

[13]. Ramanathan R, Philpott E, Duan Y, Cao G. Adoption of business analytics and impact on performance: a qualitative study in retail. Production Planning & Control. 2017 Sep 10;28(11-12):985-98.

[14]. Grandon EE, Pearson JM. Electronic commerce adoption: an empirical study of small and medium US businesses. Information & management. 2004 Dec 1;42(1):197-216.

[15]. Pinzone M, Fantini P, Perini S, Garavaglia S, Taisch M, Miragliotta G. Jobs and skills in Industry 4.0: An exploratory research. InIFIP International Conference on Advances in Production Management Systems 2017 Sep 3 (pp. 282-288). Springer, Cham.

[16]. Wolf M, Kleindienst M, Ramsauer C, Zierler C, Winter E. CURRENT AND FUTURE INDUSTRIAL CHALLENGES: DEMOGRAPHIC CHANGE AND MEASURES FOR ELDERLY WORKERS IN INDUSTRY 4.0. Annals of the Faculty of Engineering Hunedoara-international journal of engineering. 2018 Feb 1;16(1).

[17]. Podsakoff PM, MacKenzie SB, Lee JY, Podsakoff NP. Common method biases in behavioral research: a critical review of the literature and recommended remedies. Journal of applied psychology. 2003 Oct;88(5):879.

[18]. Hair Jr JF, Sarstedt M. Factors versus composites: guidelines for choosing the right structural equation modeling method. Project Management Journal. 2019 Dec;50(6):619-24.

[19]. Hooper D, Coughlan J, Mullen M. Evaluating model fit: a synthesis of the structural equation modelling literature. In7th European Conference on research methodology for business and management studies 2008 Sep 25 (pp. 195-200).

[20]. Hu LT, Bentler PM. Fit indices in covariance structure modeling: Sensitivity to underparameterized model misspecification. Psychological methods. 1998 Dec;3(4):424.

[21]. NE MW, Cudeck R. Alternative ways of assessing model fit. Testing structural equation models. 1993;154:136.

IOP Conf. Series: Materials Science and Engineering 1049 (2021) 012018 doi:10.1088/1757-899X/1049/1/012018

Side-channel Analysis using Deep Learning on Hardware Trojans

J. Kokila
National Institute of Technology, Tiruchirappalli, Tamil Nadu, INDIA
E-mail: jk.cse09@gmail.com

S. Chithra
National Institute of Technology, Tiruchirappalli, Tamil Nadu, INDIA
E-mail: chithratheertham@gmail.com

N. Ramasubramanian
National Institute of Technology, Tiruchirappalli, Tamil Nadu, INDIA
E-mail: nrs@nitt.edu

Abstract. A power side-channel analysis is proposed to nd the correct key in hardware trojan affected AES by evaluating correlation between every sub-key guesses. A deep learning method for side-channel analysis (SCA) is proposed, which consists of two phases-characterization and attack. Multilayer perceptron (MLP) and convolutional neural networks(CNN) are used as the models to determine the SCA and their performances are evaluated. Two kinds of desynchronization were used, where maximum possible value with which the trace can be shifted was 0 in rst and 100 in the second. SCA is performed in these desynchronized traces also.

1. Introduction

Side channels are those values whose analysis will help us to find some secret information. Some side-channel analysis procedures require the details about the internal operation of the system and some others do not. One of the most important usages of SCA is in the detection of hardware trojans. So there are many SCA based methods used for the same. The electromagnetic modeling about the spectrum using side-channel methods was discussed without the use of the golden chips or layouts [1]. This method needs an activated trojan to detect it. The method discussed in [2] also detects hardware trojans on AES circuits. Initially, on a normal FPGA, the AES circuit was running after which a trojan's leaked information is added. By using some instruments and adjusting the behavior of trojans, the SCA is performed to find the abnormality. In [3], SCA is used in the generation of keys based on biometrics, rather than finding the encryption key, it tries to find the signature of the user. In [4], a cross-device attack using deep learning is discussed on an AVR microcontroller. It takes into consideration, the variation within different devices by making a classifier using 30 devices in two groups. An SCA method [5] on the RSA algorithm was discussed. The data exploited were the power traces and electromagnetic radiation. The profiling attack on RSA algorithm considering the practical issues like preprocessing of observations and acquisitions. A solution for some prevention techniques against side-channel attacks by creating some misalignment in traces is dealt with in [6]. It proposes a strategy based on CNN which is a replacement for the state of the art attacks, which will include the pre-processing details in the learning phase. The side-channel attack resistance was investigated of various FPGA hardware implementations of the ARIA block cipher in [7]. It showed that the unprotected implementation of the ARIA leads to the recovery of the secret key with the low power measurements.

In this article, the main SCA work that has been done is based on the power for detected hardware trojan. From the obtained traces, the change in the consumption of the power is used to analyze and to thereby get the secret information. The traces that are used for the SCA was collected while encrypting the plain-text. Since the deep learning model is used, the obtained data are used as the basis with which the model is made and trained by using the MLP and CNN described below.

2. Proposed Model- Side-channel Analysis using Deep Learning

2.1. Design Parameter for Side-channel Analysis

In any cryptographic algorithm, if we insert any hardware trojan or any external unwanted circuit, it will make it easy to monitor the changes in the design parameters such as power, path delay, etc [8]. Such vulnerable design parameters are known as side channels. Upon monitoring the values of these side channels with the help of various inputs, it is possible to find some secret information from the cryptographic model. The method of power analysis that has been used in this project was correlation power analysis. In this, we will have an assumed power consumption model which will be used to have a key guess and compare the correlation between the assumed and actual power consumption model. If a correlation exists, that means the guessed key is right.

2.2. Correlation Power Analysis

Initially, while encrypting the plain text, actual power values will be collected. Now, a new power consumption model is assumed. The simplest and most commonly used power consumption model is the Hamming Distance Model. Hamming distance between two binary numbers, x and y is the number of different bits. The attack happens at a particular point x in the encryption algorithm. If that variable has been changed to y by the victim, we calculate the hamming distance between x and y, if the power consumption model is the Hamming distance model. The attack is done sub-key wise. We will have a key guess. For each key guess and the known plain text, we calculate the power consumption according to the guessed model. Then, the correlation coefficient is found between the assumed and actual power consumption and finds the key which correlates the best. The same procedure is done for every 16 subkeys to find the entire key value. Initially, the power traces are collected from the system while encrypting the plain-text. After that, a power model is assumed by the adversary. With the help of the assumed model and a key guess performed by the adversary, an intermediate value is found out. The correlation coefficient of the intermediate value and the normal trace is found. The process of comparing the correlation between the original and assumed model is found out for every key guess and the one that best correlates with the actual model will be the correct key. Since the key-space of the AES circuit is very vast, the exploration space is divided into 16 subkeys each size is 1 byte long. So for getting the best correlation of the key guess, it has to try for every 256 possibilities that too 16 times to get the entire key.

2.3. Side Channel Analysis using Deep Learning

The profiling SCA has two phases - the characterization phase and the attacking phase. We can consider these phases similar to the training and testing phase. In the characterization phase, the attacker has to construct a generative model that estimates a probability value each

IOP Conf. Series: Materials Science and Engineering 1049 (2021) 012018 doi:10.1088/1757-899X/1049/1/012018

for every possible key value. When we are doing a known plain text attack, for a plain text, and a guessable key chunk, we will design a probability function, given those values what is the probability that the leakage variable L is equal to the given value 1. This model will be created for every possible key value. So the generative model g(k) which is made for every possible key value, for the given leakage 1 and plain text p. In the attack phase, the attacker has to correctly predict the key value. In this, the attacker gets a new attack set in which (1,p) is there, and the key k* is fixed but unknown. From the generative model, the attacker finds the model which most likely knows the attack set. A maximum-likelihood strategy by estimating the likelihood for every candidate is found. For a subkey j, we take n sets of power traces and plain texts and outputs a distinguishing vector D which assigns a score for each possible subkey under consideration. With the help of the scores, the model will find the appropriate generative model and predict the key.

2.4. DataSet Used

The dataset will contain the traces that were collected while encrypting the plain-text in Zynq Board using an oscilloscope. This dataset is divided into two sets. One set, just like the training dataset, is needed to create the generative model (profiling model) as mentioned earlier. That set is known as profiling traces. The other one is the set of attacking traces used to retrieve the correct key from the available traces. Once the dataset is divided, it has to create a profile model. This model is created using neural networks. MLP and CNN are used to do the SCA and the results are compared [9]. After the model has been created using the set of profiling traces, the SCA is performed on the set of attacking traces to find the correct key guess.

3. Implementation Details

The hardware, software requirements along with the hyperparameter values for the proposed model is listed in Table 1.

Table 1. Hardware , Software Requirments with hyperparameter values

System Configuration		Software Requirments		Hyperparameter	Value
Processor	3.4GHz Octacore	Language	Python	Optimizer	RMSProp
Memory	8GB RAM			Learning rate	0.00001
Operating System	Ubuntu 16.04	Libraries	Keras, Tensorflow, Sklearn, Matplot	Weight Initialization	Random
				Activation function	ReLU, softmax
Dataset	Zynq Board			Loss	Categorical cross-entropy
				Number of hidden layers	4

The dataset was composed of traces and the metadata. This is stored in a hierarchical data format. The metadata in HDF5 is a compound dataset which is similar to a struct in the C language. It will contain the mask, plain-text, cipher-text, etc. This dataset is divided into profiling set and attack set. 50,000 traces are used as profiling set and 10,000 traces are used for attacking set. There are three types of datasets. One is the same thing that has been mentioned before. In this, the traces are synchronized without any jitter. Another two datasets, with desynchronization factors of 0 and 100 respectively. Table 1 shows the hyperparameters that were used for tuning the MLP model. The training set, that is the profiling set, which contains 50,000 tuples, 45,000 tuples were used for the learning purpose and the 5,000 were used for validating the model. Then another 10,000 tuples were used as the test split. This problem of finding the correct key can be treated as a classification problem. That is, prediction of the key value of AES circuit completely in one go is difficult. So we split the keyspace into 16 parts, each part having a size of 1 byte. So the possibility of correct sub-key in one round

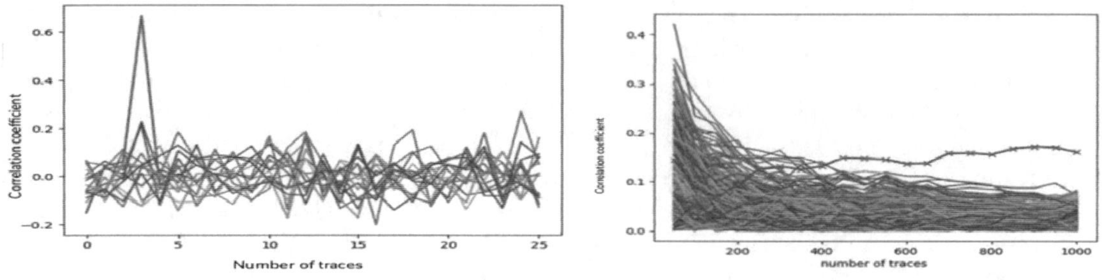

Figure 1. Key guess for one byte and correlation coefficient for all the values

is 256. So we have to classify a given trace into any of the possible 256 classes. So that is the reason why we are using categorical cross-entropy as the loss function. Regarding the CNN model, most of the hyperparameter values that have been used are the same. There are 5 pooling layers used on CNN. The optimizer used is RMSProp, which normalizes the gradient by keeping a moving average of root mean square values. An activation function is certain standard mathematical equations that help in finding the output of a neural network [10]. This function will be associated with each neuron in the network and determines if it has to be activated, based on the input given by each neuron is relevant for predicting the final output correctly. Activation functions also help normalize the output of each neuron to a range. The activation functions used in this project are reLU and softmax. ReLU is also known as a rectified linear unit will give an output zero if the input is less than zero and will give output one, otherwise. It has better gradient propagation, efficient computation, and is scale-invariant. In the final layer, we use softmax, because it will give a value between 0 and 1, and is, therefore, probability distribution.

3.1. Feed-forward Neural Network

The weights on each edge are initialized into a random value at first. Then each layer will feed its output to the next layer as its input and final output is predicted. This is known as a feed-forward network because the flow of information takes place in the forward direction, as x is used to calculate some intermediate function in the hidden layer which in turn is used to calculate y. In this, after the neural network predicts an output, the weights of each link are adjusted so that the error rate is reduced and the model will make a better prediction in the next run. For each trace, we are shifting the trace by a random number x [7]. This random number can be chosen between a value of [0..N], where N has been chosen as 50 and 100 in the two data sets that have been used. This is used to find the efficiency of the model. After we find a random value x for every trace, we shift the trace to the left In this way, 2 new data sets are created from the original one. Once the data set is created, we train for the MLP and CNN models and run on the attack set. Finally, the rank of the output is compared to confirm if the key that has been found is correct or not. The dataset with desynchronized traces helps to find which of the neural network model outperforms the other in the presence of jitters.

4. Results and Discussion

The figures show the results of the power analysis. Figure 1 and figure 2 show the key guess for one byte and the correlation coefficient for all the values respectively. It is plotted between the correlation coefficient and the number of traces. For each key guess, the correlation coefficient is shown, the one for which it shows a peak coefficient is a right key. This shows the 16 correlation power analysis retrieving all the key bytes by showing the progression of each attack depending

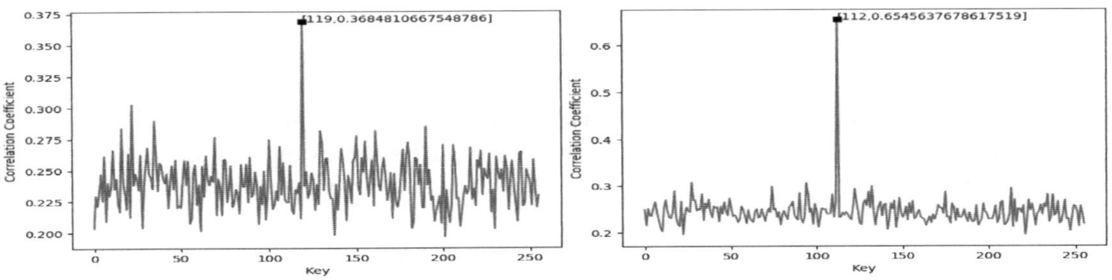

Figure 2. Correlation coefficient of a power trace for possible key values

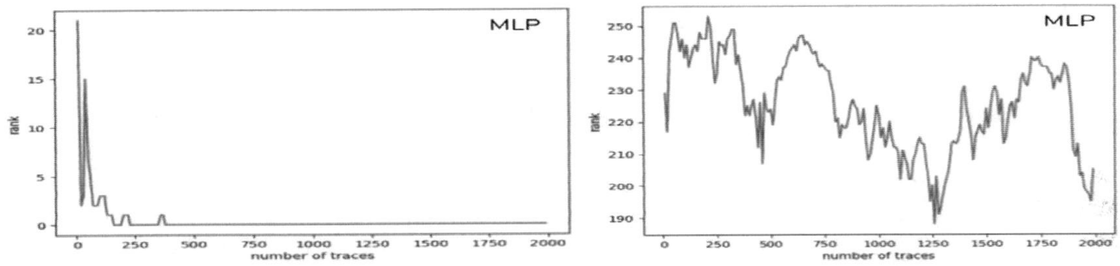

Figure 3. MLP with no desynchronization and MLP with desynchronization value of 100

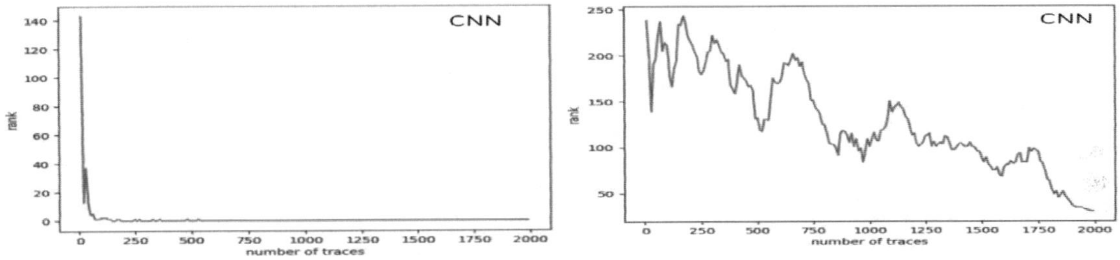

Figure 4. CNN with no desynchronization and CNN with desynchronization value of 100

on the number of traces. Figure 2. shows the second-order correlation power analysis to retrieve AES key byte. The second-order means that the sensitive data is masked. A centered product processing method is used to recombine the samples and finally, classical CPA is used to find the key byte. In AES-128, the size of the key is 128 bits. So, the keyspace will be in the order of 2128, which is very high. In such a case, the guessing of the key from that big range is extremely difficult. So, to make the key guess possible, we divide the entire key into 16 subkeys each of size 8 bits. So, the sub keyspace is reduced to 256, which is computable. So, in the following, each sub-key byte is calculated. In figure 2, we are drawing the correlation coefficient of a power trace concerning every possible key value, and it is found that for a particular key, it is showing a very high correlation factor which shows that it is a correct key.

The evaluation metric used is the rank. For every key guess, a hypothesis function is run. The key which gets the highest value will be considered as a candidate key. Now, the rank of the same is measured. That is the number of values that give the value greater than the

IOP Conf. Series: Materials Science and Engineering 1049 (2021) 012018 doi:10.1088/1757-899X/1049/1/012018

hypothesis which means if the key guess was right, then the candidate key will have a rank zero. The results obtained when the model is trained with CNN and MLP are shown in figures 3 and figure 4 respectively. We can see that in the case of without desynchronization if we compare the MLP and CNN, the rank of the candidate key is approaching zero with the number of traces around 100 on CNN. But, in the case of MLP, although the rank is getting converged to zero at a point, it takes around 400 traces. So, we can see that CNN outperforms MLP in the case without desynchronization. In the case of desynchronization, where a random shift of the traces is made, to allow jitter, both the methods are taking more time to converge. In those cases also CNN is found to be better. In figure 3, we can see that the MLP is not getting converged to a rank of zero, but we can see that in figure 4, the rank is getting converged to zero, because of which we can conclude that CNN was the best option among the two models.

5. Conclusion

The performance of SCA is evaluated using deep learning in the presence of jitter, in which two datasets with different levels of de-synchronizations were tried and their rank is compared using both the MLP and CNN models. The desynchronization was added up to a maximum of 50 in one dataset and 100 in the next. The traces are shifted left up to a random value and thus create the new set of data. In this, CNN was found to be converging to a rank zero with the given traces and found to be giving a better performance as compared to MLP. Thus, to conclude, CNN was the better choice. This work can be extended by trying for other neural network or machine learning models. Also, this can be extended to work on an extended keyspace of AES cryptography. Also, for masked implementation of the AES, the work can be tried out.

References

[1] J. He, Y. Zhao, X. Guo, and Y. Jin. 2017 . Hardware trojan detection through chip free electromagnet-icside-channel statistic alanalysis. IEEETransactionsonVery Large Scale Integration (VLSI) Systems, 25(10): 29392948.

[2] B. Hanindhito and Y. Kurniawan. 2019. Hardware trojan design and its detection using side-channel analysis on cryptographic hardware aes implemented on fpga. In 2019 International Conference on Electrical Engineering and Informatics (ICEEI) :191196.

[3] D. Delivasilis and S. Katsikas,. 2006. Side channel analysis on biometric-based key generation algorithms on resource constrained devices. International Journal of Network Security 3(1).

[4] A. Golder, D. Das, J. Danial, S. Ghosh, S. Sen, and A. Raychowdhury. 2019. Practical approaches toward deep-learning-based cross-device power side-channelattack IEEE Transactionson Very LargeScale Integration(VLSI) Systems 27(12): 27202733.

[5] M. Carbone, V. Conin, and M. Cornelie. 2019. Deep learning to evaluate secure rsa implementtions. IACR Cryptology ePrint Archive.

[6] E. Prouff, E. Cagli, and C. Dumas. 2017. Convolutional neural net-works with data augmentation against jitter-based countermeasures -proling at-tacks without pre processing. in Cryptographic Hardware and Embedded Systems - CHES 2017 19th International Conference: 4568.

[7] M. S. ChangKyun Kim and S. Moon. 2008. Differential side channel analysis attacks on fpga imple-mentations of aria. in ETRI journal 30(2): 315325.

[8] A.Kulkarni,Y.Pino,andT.Mohsenin. 2016. Adaptive real time Trojan detection framework through machine learning, in 2016 IEEE International Symposiumon Hardware Oriented Security and Trust (HOST), : 120123

[9] R. Benadjila, E. Prouff, R. Strullu, E. Cagli, and C. Dumas. 2018. Study of deep learning techniques for side-channel analysis and introduction to ascad database in Cryptology ePrint Archive.

[10] L.T.Vanschoren. 2019. Openml: networkedscienceinmachinelearning. In SIGKDD Explorations, vol. 15.

IOP Conf. Series: Materials Science and Engineering 1049 (2021) 012019 doi:10.1088/1757-899X/1049/1/012019

Modified direct clustering algorithm for group formation in cellular manufacturing

G C Potdar[1], S U Sapkal[1] and A S Shivade[1]

[1]Department of Mechanical Engineering, Walchand College of Engineering, Sangli, Maharashtra, India.

potdar.c.gaurav@gmail.com, sagarus1201@gmail.com, shivadeanand@yahoo.in

Abstract. In cellular manufacturing system formation of machine-part families is an important task. Cellular manufacturing systems deals with various clustering algorithms which are particularly relevant to the problem of machine component group formation. In this paper, case study from well-known gearbox manufacturing company is considered where there is scope for using cellular manufacturing system. Company is manufacturing the gearbox varying from single stage reduction to four stage reduction. A modified direct clustering algorithm is proposed and is applied after the initial incidence matrix derived from direct clustering algorithm. This algorithm overcomes the problem arises during application of direct clustering algorithm and gives the optimum solution. The algorithm is specifically designed to deal with large amount of data during realistic situations. The results of proposed algorithm are compared with well-known existing algorithms and it is found that the proposed algorithm gives the better solution than the existing algorithms under consideration.

1. Introduction

Group technology deals with the fragmentation of a manufacturing system into subsystems [4]. In order to enable this fragmentation various coding systems and classification have been developed. With the help of these codes, clustering of machine-component into subsystems is accomplished. Clustering analysis is one of the most frequently applied mathematical tools in group technology [6]. Cellular manufacturing is an application of GT to manufacturing. In cellular manufacturing, a few machines, usually dissimilar in function, are grouped into a cell. For mass production a dedicated cell is required which enable to process a family of parts [6].

An XYZ Pvt. Ltd. company which produces gearbox is being considered as a part of work. Company has made changes in their plant layout earlier but failed to meet desired objectives. Distances in between the processes is not feasible as far as plant layout is concern. To overcome these problems company has decided to use cellular manufacturing system. Presently, gearbox varying from single to four stage reduction is being manufactured in the plant. There are common processes and common components in between single to four stage reduction gearbox. Consider an example of single stage gearbox having 13 components and 23 processes. Some processes are common in those 13 components. So, clustering those common processes by applying various algorithms is the requisite task.

As described by Arumugam S [9] the input for a cellular manufacturing problem consists of , a set Y of p parts and a set X of m machines and an m × p matrix $A = (a_{ij})$, where $a_{ij} = 1$ or 0 according as the part p_j which is processed on the machine m_i. In the context of similarity, it includes design attributes such as size, shape, etc. and/or manufacturing attributes like length, diameter, surface finish, tolerance, etc. Once if the parts are grouped, then machines, tools and equipment required to process these parts having similarity are grouped together to take the advantages of similarities in manufacturing.

Burbidge J L [1] described that there are three stages in the analysis of the production flow in which the first stage is to determine the design of the department in the factory through the analysis of the factory flow, the second stage is to determine the group of machine components in the group analysis and the third stage is to determine the layout of each group of machine components in the line analysis. Burbidge J L [2] suggested a method for sorting components using punched card processes. This technique created a flow sequence based on the map, but it failed to define a grouping. King J R [3] reviewed the single cluster analysis method, the production flow analysis method, as well as the bond energy method, and built a rank cluster method. Kusiak A [4] developed an integer programming model that offers a more convenient representation of clustering problems than a matrix model that enhances the performance of component families and machine cells. Ballakur A et al [5] developed a heuristic focused on cell utilization for the design of cell manufacturing systems. Machines are allocated to cells based on workload and cell size limitations. Components are 'assigned' to cells in such a way that most of their operations can be done within cells. Seifoddini H et al [6] analyzed the components of the machine cell formation cycle such as Similarity coefficients, clustering algorithms, and efficiency measurements. New consistency metrics are implemented and a comparative analysis of three different similarity coefficients is performed -the Jaccard similarity coefficient, the weighted similarity coefficient, and the commonality value. Alhourani F [7] Implemented similarity coefficient that incorporates multiple process routing along with operation sequence, production volumes, duplicate machines, and machine capacity. They also proposed a clustering algorithm for the formation of machine cells. The developed similarity coefficient showed greater sensitivity to intercellular movements and improved machine grouping. Telsang M et al [8] implemented cellular manufacturing system to reduce intercellular movements, transport waste, employees' wages to increase productivity. Arumugam S et al [9] developed an algorithm for the identification of bottleneck machines and bottleneck sections in a cellular manufacturing problem along with a cell partition. Mehta S et al [10] developed differential bond energy algorithm and results are compared to the classic bond energy algorithm. Dhayef D et al [11] developed branch and bound algorithm for machine cell and part family and for sequencing and scheduling groups for makespan calculation. The developed algorithm was evaluated by a case study consisting of four items processed on nine machines. Built algorithm can be used to decide the best job and minimize the order of the best group for any kind of problem and completion time (makespan). Chan H M et al [12] developed a direct clustering algorithm that shapes component families and machine groups for cellular manufacturing by iteratively restructuring the machine component matrix. The direct clustering algorithm consists of the matrix being concatenated, moving the rows with the leftmost assigned to the top and the columns with the topmost assigned cell to the left of the matrix.

2. Direct clustering algorithm

2.1. Introduction

Direct clustering algorithm is the method of finding machine and component groups by revamping the sequence in machines and components which are listed in the matrix. The initial incidence matrix is constructed by allocating '1' to the cells when a component goes through a process in a particular machine and '0' when it is not. While applying clustering algorithm it is desired to form a diagonal clustering of component and machines. Therefore, the matrix is sorted in two stages. Firstly, sorting is being concluded based on the counts of the number of entries of '1', columns and rows are sorted in decreasing order of counts. The predicament emerges in second sorting where it postulates the sort as 'leftmost' rows allotments to the top and 'topmost' column allotment to the left. When it is attempted to iterate it for the second sorting, the diagonal element formed is not optimum and can be iterate further. Thus, the algorithm is unable to come to the final matrix which indicate diagonalized form of clustering.

The industrial data of 13 components and 23 processes is converged to 12 components and 12 processes by eliminating common allotments in all components (for example, inward inspection is common for all components so it can be eliminated) and the unique allotment for a single component (for example, inner diameter grinding is a unique process for planet gear and no other component undergone such process). It is not necessary that every time the number of components must be equal to number of processes.

2.2. The clustering algorithm

a. Count the number of '1' in each column and row. Regroup the machine component incidence matrix with rows and columns in decreasing order with virtue of counts of '1'.
b. Contemplating firstly with column number one of the matrix relocate the rows which have '1' as an entries in this column to the top of the matrix. Continual of this procedure with respect to the other column will give the new arrangement for the rows.
c. Is the current matrix being optimum diagonalized?

if yes, then go to 'f'
if no, then go to 'd'

d. Rerun the procedure mentioned in 'b' for rows, starting from row 1. This will give new arrangement for the columns.

e. Is the current matrix being optimum diagonalized?
if yes, then go to 'f'
if no, then go to 'b'

f. Stop

2.3. Application

The machine component incidence matrix with 12 machine and 12 components is explicated in Table 1. Table 1 illustrates the matrix after the first step which is used in direct clustering algorithm. A row numbering from 2 to 4 designate to components and column numbering from 6 to 7 designate to machine or process. The bottommost row and rightmost column designate the count of number of entries of '1' in each column and row respectively.

Table 1. The initial incidence matrix using direct clustering algorithm.

S. No.		1	2	3	4	5	6	7	8	9	10	11	12	Number of entries in row ⬇
		2	5	12	8	10	9	6	3	1	11	7	4	
1	6	1	1	1	1	1	1	1	1	1			1	10
2	4	1	1	1	1	1		1	1	1	1			9
3	10	1	1	1	1	1	1				1	1		8
4	9	1	1	1	1	1	1				1	1		8
5	2	1	1	1	1			1	1		1			7
6	3				1	1	1	1	1	1			1	7
7	1	1	1	1				1	1	1	1			7
8	8	1		1	1		1					1		5
9	5	1	1						1	1			1	5
10	12			1		1		1						3
11	11	1	1				1							3
12	7	1	1							1				3
Number of entries in column ➡		10	9	8	7	6	6	6	6	6	5	3	3	

In Table 1, row 6, 4,10,9 and 2 clustering with column 2,5,12 and 8. Continual of iteration through the rows and column it is observed that these formed clustered get disturbed and they get deviated from their desired results of forming machine component family through their diagonal arrangement. The iteration can be done by relocating leftmost row to the top and topmost column to the left. As this iteration process does not follow a proper path for clustering, it becomes difficult to determine optimum diagonal elements. As this diagonal element can further be optimized which can determined by using modified direct clustering algorithm. Modified direct clustering algorithm enables to allocate possible optimum cell arrangement by assigning weights based on the priority of the positioning in the diagonal form.

3. Modified direct clustering algorithm

3.1. Introduction

As discussed earlier, modified direct clustering algorithm is applied after an initial incidence matrix formed in direct clustering algorithm. The aim of clustering is to diagonally form machine component family matrix in three clusters of 4 × 4 (only for 12 × 12 data). First cluster consist of the elements allocated at upper left most part of the matrix (row and column from s no.1 to 4 in table 1), followed by second cluster which is at the middle part of the matrix (row and column from s no. 5 to 8) and third one which is at lower right most part of the matrix (row and column s no. 9 to 12).

The matrix has undergone the first sorting which is based on the counts of entries '1' in the cell. As discussed in King J R [4], weights are the most important parameter in the clustering. Weights can be given as per the element's configurations required. Here, it is required to cluster the elements in diagonal form in three clusters as mentioned above. Therefore, it is essential to give the priority to first four cells of each column and row as 4^3 and remaining eight cells as 4. Repeat this procedure with respect to middle four values and last four values of each row and column. Now the priority for three clusters of 12 rows and three clusters for 12 columns is set. Sum up those priorities for each row and column and apply the equation 1 for finding out the weights. Seifoddini H [6] worked on similarity coefficient algorithm in which Jaccard's coefficient is applied. Referring to the Jaccard's coefficient the formula is applied as mention in equation 1.

$$W_i = \frac{\text{Sum of priorities of first cluster for } i^{th} \text{ row}}{768} \tag{1}$$

W_i is the weight for i^{th} row and 768 is the maximum sum of the priority for a row or a column ($4^3 \times 12$). Similarly, equation 1 is applied for other row and columns.

3.2. The proposed clustering algorithm

a. Rearrange the rows and columns to form it as initial incidence matrix as in direct clustering algorithm (based on counts of number of entries of '1' in decreasing order)
b. Formation of the first cluster in which s no. 1 to 4 of rows and columns allotted the priority of 4^3 and for remaining eight elements priority of 4.
c. Sum up the priority for each row and column for constructing the first cluster and apply the equation 1 individually.
d. Sort according to weights in decreasing order and then relocate the rows and columns in decreasing order of their weights.
e. Identify the diagonal elements (while identifying do not repeat the element number in a cluster).
f. Is the current matrix being optimum diagonalized?
   if yes then go to 'g'
   if no then go to 'e'
g. Repeat the procedure 'b', 'c' and 'd' for second cluster and third cluster.
h. Is the current matrix being optimum diagonalized?
   if yes then go to 'k'
   if no then go to 'e'
i. Check for other diagonal elements by rearranging rows and columns (keep on eliminating allotted or overlapped elements)
j. Is the current matrix being optimum diagonalized?
   if yes then go to 'k'
   if no then go to 'e'
k. Stop

The steps in the proposed algorithm are applied on 12×12 incidence matrix. The machine component incidence matrix using modified direct clustering algorithm is illustrated in table 2.

Table 2. The machine component incidence matrix using modified direct clustering algorithm.

S. No.		1	2	3	4	5	6	7	8	9	10	11	12	Number of entries in row
		2	5	12	8	6	3	10	9	1	11	7	4	
1	6	1	1	1	1	1	1	1	1	1			1	10
2	4	1	1	1	1	1	1	1		1	1			9
3	10	1	1	1	1			1	1		1	1		8
4	9	1	1	1	1			1	1		1	1		8
5	2	1	1	1	1	1	1				1			7
6	1	1	1	1		1	1			1	1			7
7	3				1	1	1	1	1	1			1	7
8	8	1		1	1				1			1		5
9	5	1	1				1			1			1	5
10	12			1		1		1						3
11	11	1	1						1					3
12	7	1	1							1				3
Number of entries in column		10	9	8	7	6	6	6	6	6	5	3	3	

4. Results and discussion

In modified direct clustering algorithm, there are two stages of sorting which are used after the first sorting from direct clustering algorithm. The second and third sorting from the proposed algorithm concentrates more on diagonal configuration of the matrix. In second sorting, priorities are divided in clusters of three (s no. 1 to 4, s no. 4 to 8 and s no. 9 to 12 for both rows and columns one by one). Priorities are assigned only to cells which has entry as '1'. Firstly, the priority of 4^3 is given for the first cluster (s no. 1 to 4) and for remaining it is 4. As mentioned by King J R [4], the priorities should be given in terms of exponent. For example, if 4, 3 and 2 are the priority for first cluster, second cluster and third cluster respectively. Then the purpose of diagonal configuration of the matrix will not be solved because if number of counts of '1' increased in second and third cluster then weights (which is the part of third stage) obtained will tend towards second and third cluster. This will form element in row number s no. 1 to 4 and column number s no. 5 to 8. In third sorting, the weights are given based on the priority for machine and component and they are arranged in the descending order as shown in table 2. From table 2 it is observed that the elements are diagonalized and there is no scope for the further optimization. After finding out the clusters, elimination of overlapping elements becomes essential as it may affect in bottlenecks. The remaining rows and columns are rearranged manually and configured diagonally as mentioned in the table 2. Thus, the optimum solution is achieved and it is considered as final solution.

Contemplating from table 2 that machines 6,4,10 and 9 forming a cluster with components 2,5,12 and 8. On the other hand, machines 2 and 1 forming a cluster with components 6 and 3. Similarly 3-10, 8-9 and 5-1 are also the effective combinations. Further clustering is not possible as it may lead to the construction of bottlenecks.

5. Conclusion

The optimum solution for system component clustering is not feasible in the case of a direct clustering algorithm, because there is no efficient way to perform iterations. This algorithm is then modified by constructing a machine-component family with a diagonal structure and called as a modified direct clustering algorithm. A comparative evaluation of the machine component incidence matrix is carried out and it is found that direct clustering algorithms are less efficient when it comes to realistic problems as mentioned in earlier research. Moreover, it is found that modified direct clustering algorithms can handle large data and practical problems as they deal with priority by using the initial incidence matrix. Bottlenecks can also be found effectively by a modified direct cluster algorithm.

References

[1] Burbidge J L 1963 Production flow analysis *The Production Engineer* **43** p 742.
[2] Burbidge J L 1969 Production Flow Analysis *Int. Centre for Advanced Technical and Vocational Training*
[3] King J R 1980 Machine-component grouping in production flow analysis: an approach using a rank order clustering algorithm *Int. J. of Production Research* **18(2)** pp 213-232
[4] Kusiak A 1987 The generalized group technology concept. *Int. J. on Production Research* **25(4)** pp 561-569
[5] Ballakur A and Steudels H J 1987 A within-cell utilization based heuristic for designing cellular manufacturing systems *Int. J. of Production Research* **25(5)** pp 639-666
[6] Seifoddini H and Hsu C 1991 Comparative Study of Similarity Coefficients and Clustering Algorithms in Cellular Manufacturing *J. of Manufacturing Systems* **13(2)**
[7] Alhourani F 2013 Clustering algorithm for solving group technology problem with multiple process routings *Computers & Industrial Engineering* **6** pp 781–790
[8] Bharamgude P and Telsang M 2014 Design, analysis and performance simulation of cellular manufacturing system - a case study *Int. J. of Mechanical And Production Engineering* **2(6)** pp 2320-2092
[9] Arumugam S and Saral J 2015 Identification of Bottleneck Elements in Cellular Manufacturing Problem *Int. Conf. on Graph Theory and Information Security* **74** pp 181 – 185
[10] Mehta S Agarwal P Shrivastava P and Barlawala J 2018 Differential bond energy algorithm for optimal vertical fragmentation of distributed databases *J. of Computer and Information Sciences*
[11] Dhayef D and Abdulkhafour A 2018 Developing branch and bound algorithm for cell formation and group scheduling *Engineering and Technology Journal* **36** pp 555-563
[12] Chan H M and Milner D A Direct Clustering Algorithm for Group Formation in Cellular Manufacture *J. of Manufacturing Systems* **1(1)**

Acknowledgement

The authors are grateful to the Kavitsu Transmission Pvt. Ltd., Satara for providing the data and necessary facility.

Estimating creditworthiness using profit modeling

S Ramesh

Jayachamarajendra College of Engineering, JSS TI Campus, Manasagangotri, Mysore - 570006, India

E-mail: sagarika251408@gmail.com

Abstract. The recent advancements in machine learning have enabled retail credit issuers to predict a customer's net present value using behavioral and profit-based modeling approaches. The objective of this research is to curtail the risk associated with credit card underwriting which typically helps in deciding whether an applicant is creditworthy or not. It attempts to estimate the creditworthiness of a customer instead of evaluating the chances of them going delinquent on a debt. Credit issuers can utilize these models to identify potentially risky as well as profitable customers. Using a machine learning approach, profit-based models are constructed in this research using information acquired at the time of customer-acquisition.

Keywords—Profit Models, Credit Cards, Machine Learning, Underwriting, Retail Credit Risk

1. Introduction

An underwriter is empowered to place a customer into a particular risk category upon reviewing their application. Underwriters can decline the applicant altogether as well based on their evaluation. They also rely upon extensive medical knowledge and personal experience when underwriting cases [1]. The reliance upon their own experience and judgment causes inconsistency across the underwriters, resulting in inaccurate rate classifications [1]. Over time, this process has been automated to eliminate human errors and maintain transparency in the decision-making process. Automated underwriting has been a growing trend in the retail credit industry. In recent years, consumer credit portfolios have experienced extraordinary growth in developing countries [2]. This represents huge revenue opportunities in auto loans, home loans, and credit cards, and has created a tremendous need for more efficient and cost-effective credit analysis [2].

It is important to note that risk is not the only factor for approving a customer's application. There is a possibility that a riskier customer can be more profitable than a safer customer based on the institution's product line and business strategy. Profit-based modeling is relatively rarer because of the complications that it involves but institutions have turned to it as a result of the advances in machine learning methods in credit risk modeling.

Furthermore, the relationship between profit and risk in profit-based modeling typically depends on loan type, loan life, and expenses and incurred losses in a credit card. Some notable components of card profit include sub drivers of revenue that is composed of interchange income, interest income, and fees. Interchange income is the money that a financial institution earns from swipe fees. Swipe fees are a percentage of the transaction that banks take from retailers each time a credit card is swiped to pay for a purchase [3].

For designing credit card portfolios, card issuers can rely on earn primarily on interchange income through transactors. A transactor is a consumer who pays their credit card balance in full and on time every month [4]. Transactors do not carry a balance from month to month; they always pay their credit card bills in full by the due date [4]. Card issuers can also design their portfolios based on revolvers. A revolver is a credit card issuer term for customers who carry balances, paying off those balances over time, thus "revolving" them [5].

This paper studies the empirical relationship between risk and profit pertinent to customers in order to estimate their creditworthiness. Using acquisition-level information, logistic regression model is applied to estimate the tendency of a customer to go delinquent or charge off. This estimation is used to organize accounts into interchange, neutral, and finance charge portfolios. An acquisition-level profit model is also constructed for the loans to predict credit worthiness of the customer.

Related Works

This builds on papers by Grodzicki and Koulayev [6] and Stavins [7] which study consumer credit card usage patterns generally. An influential study by Agarwal et al. [8] estimate the heterogeneous effect of credit line increases on components of marginal profit of customers across the FICO score spectrum. This component of our paper is closely related to this, but focuses on how empirical profitability relates to general measures of riskiness in cards at time of acquisition between different types of portfolios. Our paper can also be compared with the work on "Credit Risk Analysis Using Machine and Deep Learning Models" [9] that checks the decisional power of algorithms to provide loans to an enterprise. A wide variety of algorithms are studied in the research paper to single out the best one offering the highest accuracy and precision. Our paper differentiates from this work as we utilize information acquired during customer acquisition to rank customers based on risk and profitability by leveraging machine learning algorithms. The focus is directed on establishing a risk-profit relationship instead of studying numerous algorithms. Another paper that relates with our work is the study where machine learning algorithms are applied for consumer credit risk modeling [10]. A model to predict consumer delinquency is developed in the paper whereas, our paper stresses on studying customer behavior and their tendency to go delinquent taking the risk and profit landscape into account as well. Both profit and risk models are developed where the risk-model estimates the risk associated with individual loans before estimating the risk of accounts that can charge-off. For each account-level model for aspects like monthly balance, tendency to pay late fee and propensity to charge-off, hyperparameter searches and variable selection techniques are employed.

2. Methodology

2.1. Data Collection

Retail credit information is gathered from the records monitored by the Office of the Comptroller of the Currency. The data collected is composed of user credit card account activity beginning January 2008 to December 2015. It consists of actual payment information including outstanding balance, balance transfers, credit limits, among others. However, the figures acquired lack reward points that a user earns upon using a credit card. Such an inadequacy comprises the final output as it fails to account for this expenditure for an institution. It increases the likelihood of stating inaccurate profit margin for the company offering credit cards. A subset of this data set holds bureau attributes with variables relevant to trade lines. It is sourced from information gathered at the time of customer acquisition. These variables are utilized in constructing risk and ML-based profit models. Furthermore, over 100 thousand loans are analyzed by extracting data from multiple bank accounts across a window of 36 months since 2012.

2.2. Interchange, Finance Charge, and Neutral Portfolios

Information obtained from reputed banking institutions is blended with the existing data to create portfolios. These portfolios contain crucial information pertinent to a customer's chances to go delinquent, and their spending behavior. Various card issuers categorize their customers' accounts based on such premises. The factors considered while developing the portfolio algorithm that categorizes accounts on grounds of whether issuing institutions consider interchange earnings or rely on credit card finance charges are as follows: information gathered during customer acquisition, application details. Such groups are classified as 'Interchange portfolio' and 'finance charge portfolio'. There is another 'neutral portfolio' for the accounts that do not fit in the either of the two portfolios.

2.3. Risk Model

A model for estimating the risk associated with individual loans is developed before estimating the risk of accounts that can charge-off. This system sorts the loans based on their vulnerability to risky behavior and assigns points to them accordingly. Based on such points, the riskiness of a customer is predicted. The risk score is computed by leveraging information acquired during customer acquisition. It designates a loan as 'risky' if the corresponding customer is reported delinquent based on the following criteria:

a. 60 days delinquent under a short duration (6 months)
b. 90 days delinquent under a relatively longer duration (1 year)
c. 120 days delinquent under 2 years
d. Charge-off in 2 years

Over 150 thousand loans are examined by extracting data from multiple bank accounts originating in 2012 till 2015. In order to enhance the estimation, the sample is divided into two distinct parts: 20% grouped as risky and 80% non-risky and an 80-20 training validation split is made.

Logistic regression is used for risk modeling which can be described as follows:

$$log\frac{p(y=1)}{1-p(y=1)} = \alpha + \beta X + \epsilon \tag{1}$$

where $y = 1$ denotes the event that the customer is risky.

The predictors are first arranged in accordance with their univariate information value for variable selection. The variables are then segmented into decile bins and the Weight of Evidence (WOE) is computed in each of these bins which can be described as follows:

$$WOE_i = log\frac{P(i|y=0)}{P(i|y=1)} = log\frac{\frac{N(y=0,i)}{N(y=0)}}{\frac{N(y=1,i)}{N(y=1)}} \propto log\frac{P(y=0|i)}{P(y=1|i)} \tag{2}$$

IV is then calculated as a weighted sum of WOE values

$$IV = \sum_{i=1}^{m}\left(\frac{N(y=0,i)}{N(y=0)} - \frac{(N(y=1,i)}{N(y=1)}\right)WOE_i \tag{3}$$

A sample of variables are taken from interchange, finance charge, and neutral portfolios which are used to build the initial list. These are binned and after performing WOE transformations, they convert to log-odds of default in that portfolio. Now, regression is done to modify the model based on Akaike information criterion tests.

IOP Conf. Series: Materials Science and Engineering 1049 (2021) 012020 doi:10.1088/1757-899X/1049/1/012020

2.4. Profitability versus Risk - Empirical Relationship

In this section, a relationship is established to correlate profit and risk at the account level. It is presumed that the interchange income is 1% of total purchase volume.

The Net Profit Value (NPV) of each account for customer i can be described as follows:

$$NPV(i) = \sum_{t=i}^{T(i)} profit(i;t) \tag{4}$$

Here, $T(i)$ denotes the loan life of the customer i.

2.5. E. Profit Model

If a card issuer acquires customers based on ranking of profitability, the portfolio would be restricted to the middle risk area as it would lead to the addition of both high and low risk customers in uncertain proportions. However, issuers are unaware of the most profitable customers at the time of booking. They still have to decide the most profitable customer based on credit bureau attributes and application-level information such as self-reported income. As a result, the shortcomings as well as benefits of predicting a customer's profitability and risk based on information acquired during booking is observed. XGBoost is used to estimate a customer's profitability and modeling revenue, probability of default and loss given default renders the best results at the account-level. This approach failed to differentiate between the various sources of revenue so it caused difficulty with interpretation. Therefore, the revenue drivers were modeled separately and full model estimation is started after driver structure selection. For each account-level model for aspects like monthly balance, tendency to pay late fee and propensity to charge-off, hyperparameter searches and variable selection techniques are employed.

3. Results and Discussion

XGBoost package is used to build the interchange, finance charge, and neutral portfolio models with 50 trees and a 0.1 learning rate for assorting the customers. XGBoost comes with a default variable importance measure that calculates how much each attribute split point improves the performance measure, weighted by the number of observations the node is responsible for. The findings show that the interchange portfolio consists of the highest number of creditworthy customers and finance charge portfolio contain the least of them.

In tables 1, 2, and 3, the most informative variables are listed per portfolio. One can see that certain significant risk drivers are common across portfolio, such as FICO of the primary borrower at origination, total borrower bankcard utilization, and total borrower bankcard credit limit. However, for some portfolios, number, balance, or credit limit of premium cards are highly predictive of borrower risk.

Table 1. Top Predictor Information Value of Interchange Portfolio.

Rank	Variable	IV
1	FICO	1.06
2	Total Card Util.	0.72
3	Prem. Card CL	0.51
4	Prem. Card Bal.	0.48
5	Total Card Bal.	0.44

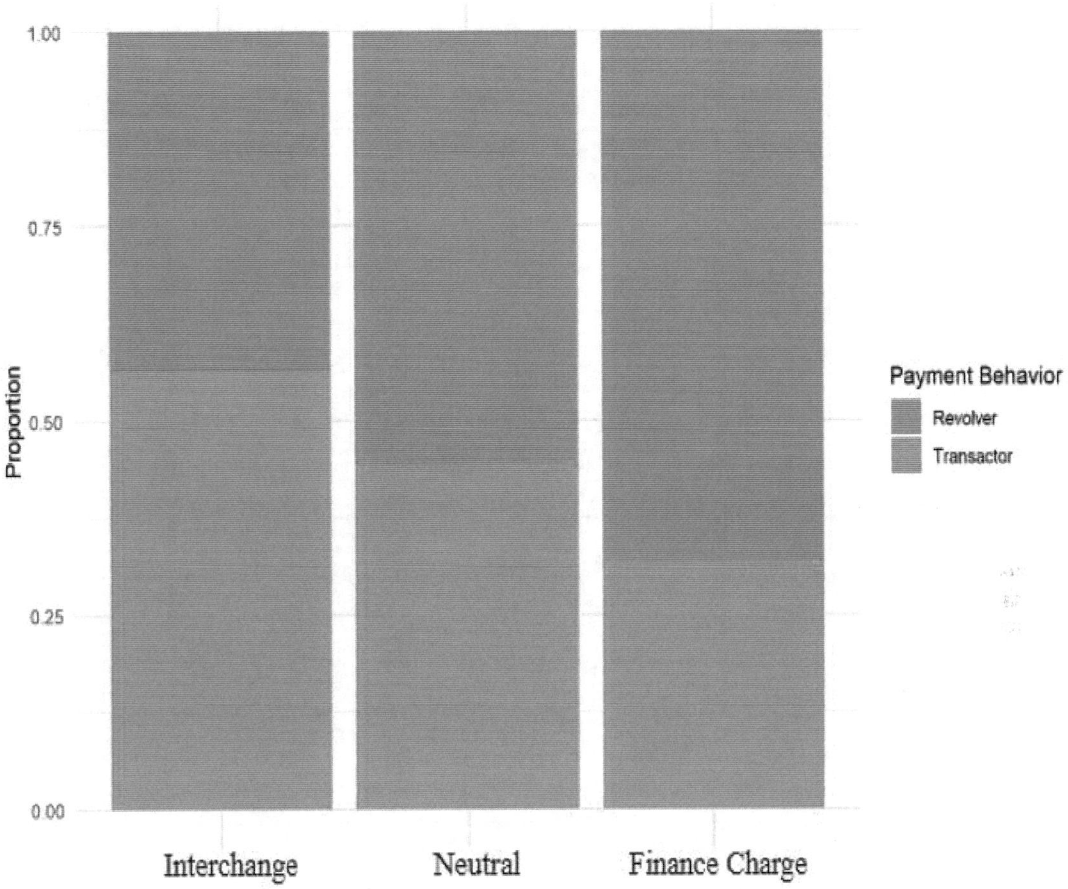

Figure 1. Transactors vs Revolvers

The maximum loan life is 36 months in this case because it accounts for data across 36 months from 2012 to 2015.

The above finance charges vs risk graphs across interchange, neutral, and finance charge portfolios are analyzed to estimate the total finance charges accumulated over the life of the loan. Low risk individuals tend to be lesser beneficial in terms of revenue generation because

Table 2. Top Predictor Information Value of Neutral Portfolio

Rank	Variable	IV
1	FICO	1.02
2	Total Card Util.	0.71
3	Prem. Card CL	0.45
4	Prem. Card Bal.	0.44
5	Total Card Bal.	0.44

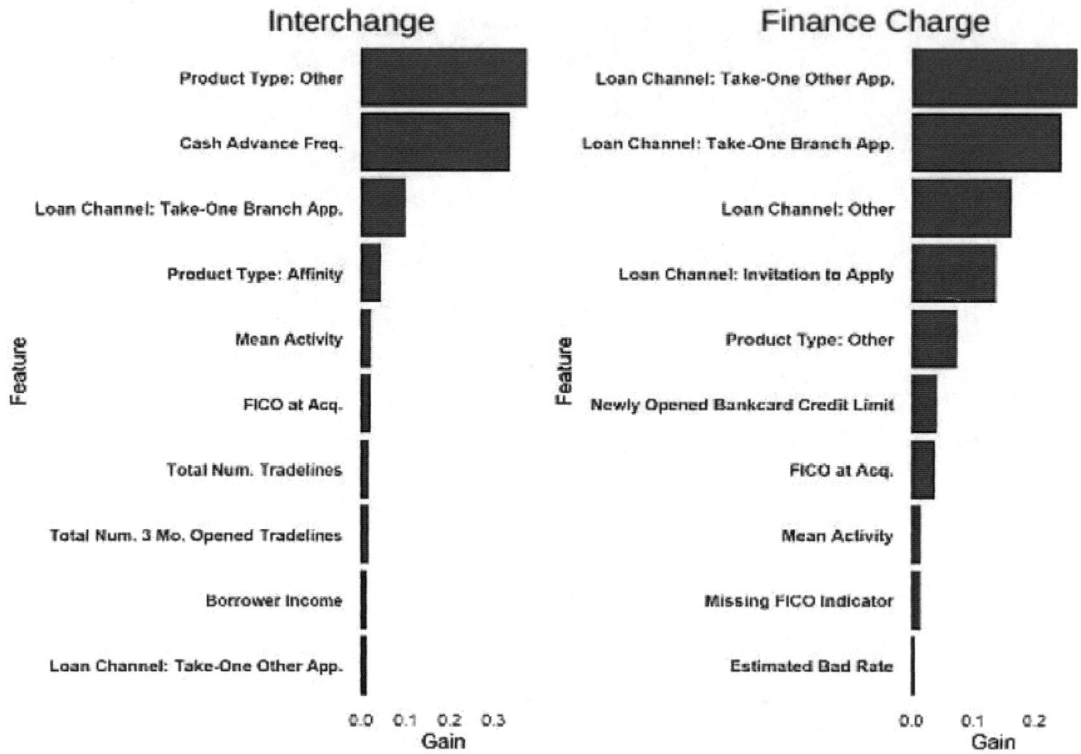

Figure 2. Portfolio Algorithm Feature Importance

they are generally transactors and do not accumulate interest. The high-risk individuals cannot accumulate large amount of interest because their credit limits are shorter and so is the duration of loan life. However, the graph indicates a point which contains enough revolvers that accumulate significant finance chargers owing to large credit limits and loan lives.

The next area of focus is interchange income. It is observed that low risk customers are less likely to use the credit card spite of having the tendency to spend more as a result of their higher income. The reason is that they have multiple cards competing with each other in their wallets. Also, the customers with minimal risk are seen to spend less and exercise control over their expenditure. Late fees and fees such as balance transfer fees, annual membership fees, and

Table 3. Top Predictor Information Value of Finance Portfolio

Rank	Variable	IV
1	FICO	0.82
2	Total Card Util.	0.46
3	Prem. Card CL	0.41
4	Prem. Card Bal.	0.36
5	Total Card Bal.	0.36

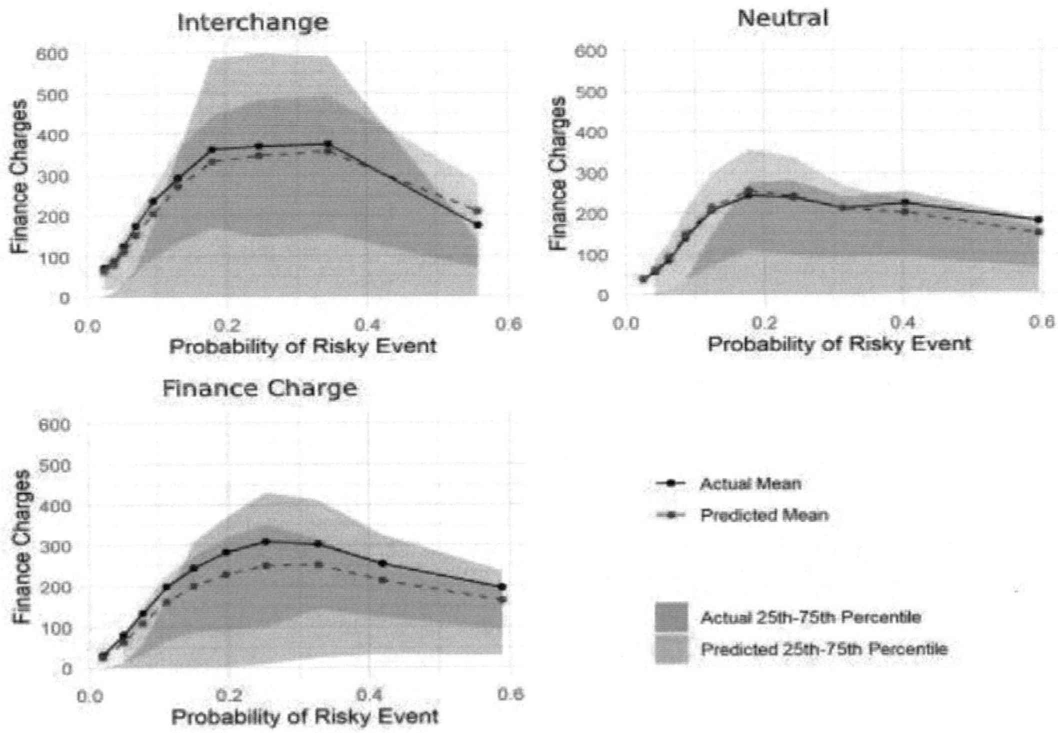

Figure 3. Predicted versus Actual Finance Charges Over Risk Spectrum

Table 4. Values depicting Predicted vs Actual Finance Charges over Risk Spectrum

Portfolio	Somer's D	R'Score
Interchange	63.8	25.6
Neutral	61.3	24.9
Finance Charge	60.0	32.8

cash advance fee might be small but they also play a key role in computing the overall profit. Although the interchange portfolio would by itself have a lower optimal revenue" sweet spot," as compared to finance charge portfolio it is also observable that an increase in risk is directly proportional to an increase in the other small fees. All such aspects combine to form the overall revenue strategy.

A grid search is performed to tune the tree depth and learning hyperparameters. The results of these searches can be summarized in the table 7 below:

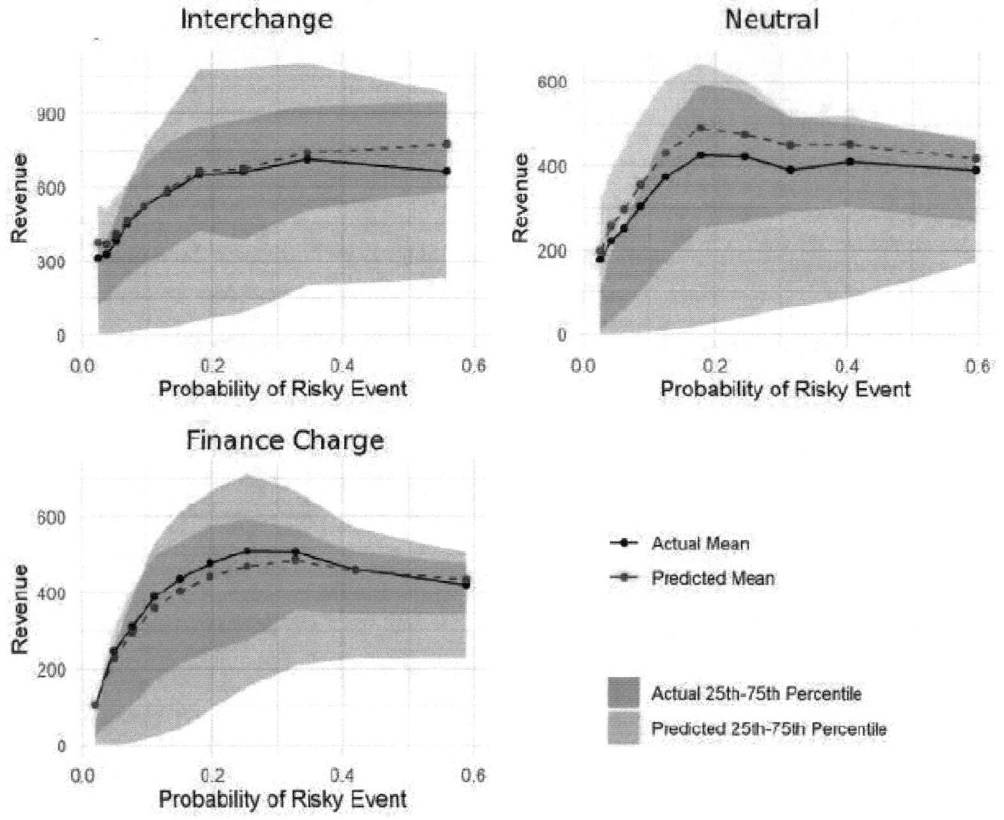

Figure 4. Predicted versus Actual Interchange Income Over Risk Spectrum

Table 5. Values depicting Predicted vs Actual Interchange Income over Risk Spectrum

Portfolio	Somer's D	R^2Score
Interchange	57.5	26.4
Neutral	63.3	29.0
Finance Charge	59.6	26.8

Table 6. Values depicting Predicted vs Actual Revenue over Risk Spectrum

Portfolio	Somer's D	R^2Score
Interchange	57.5	26.4
Neutral	74.2	29.2
Finance Charge	75.4	30.7

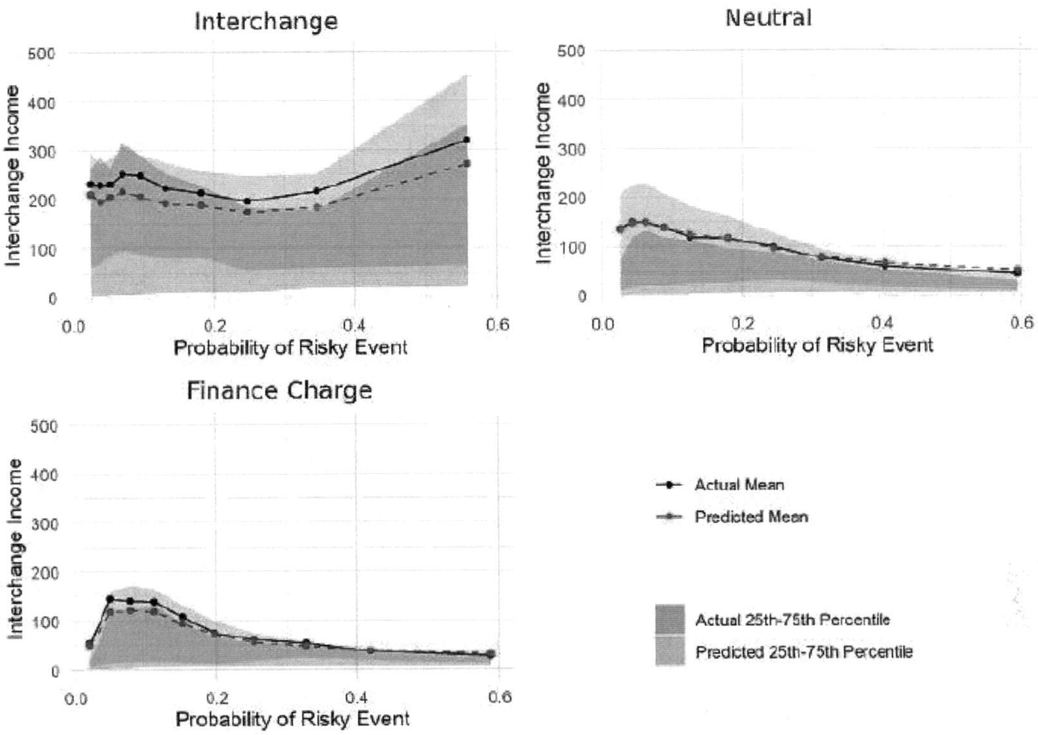

Figure 5. Predicted versus Actual Revenue Over Risk Spectrum

Table 7. Hyperparameters for Driver Submodels

Submodel	η	Tree depth	Num Trees
Finance Charge	0.009	6	2135
Interchange Income	0.001	2	6793
Late Fees	0.005	8	2056
Other Fees	0.014	7	928
PD	0.005	9	455
LGD	0.004	5	2696

Calculated on 20% holdout set.

The findings suggest that both the revenue model and aggregate profit model have poor predictive performance, but better rank-ordering performance. Profitable customers can be identified well enough with the data available at booking, as can be evidenced by the models' Somer's D across driver, in the 40 – 50 range. In addition, the hump-shaped correlation between the elements of profit and risk is maintained between predicted and actual curves. It can be inferred that a ranking based on this score is a valid for estimating customer's profitability.

IOP Conf. Series: Materials Science and Engineering 1049 (2021) 012020 doi:10.1088/1757-899X/1049/1/012020

4. Conclusion

Behavioral modeling and profit scores for underwriting is analyzed and a correlation is established between empirical profit and risk through a model developed using acquisition-level information. Financial institutions can potentially adopt these profit-based models for underwriting and loan management purposes to more precisely target profitable but potentially risky consumers. Hyperparameter searches and variable selection techniques are used for full model estimation and it is observed the results maintain the empirical profit-risk relationship. In the end, revolvers are found to be the most profitable customers with a higher income range and larger accumulated interest. The proportion of risk is higher than profit in portfolios that target low revolve rate spenders and those that are lower in the credit spectrum. Therefore, it is important to be mindful of risk-based guardrails when applying new modeling methodologies for risky customers. Portfolios that have a customer and product composition that targets low revolve-rate spenders have a minor increase in riskiness when switching to a profit score, while portfolios that are lower in the credit spectrum have a large increase in risk when using the profit score relative to a risk score. The increase in losses when using these scores is proportionally higher than the increase in profit, again concentrating in portfolios at the low end of the credit spectrum. This reinforces the well-known importance of risk-based guardrails when using new modeling methodologies in underwriting for issuers concentrating in higher risk customers.

Acknowledgments

I would like to express my appreciation to Mr. Suraj Kumar Jana for his valuable and constructive suggestions during the planning and development of this research work. His willingness to give his time so generously has been very much appreciated.

References

[1] Aggour KS, Bonissone PP, Cheetham WE and Messmer RP 2006 Automating the underwriting of insurance applications *AI Mag.* **27** 36–36
[2] Abrahams CR and Zhang M 2008 A Hybrid Consumer Credit Underwriting Decision System *ResearchGate* 1-1
[3] National Retail Federation 2020 *Swipe Fees* `https://nrf.com/swipe-fees`
[4] Chen J 2020 Transactor *Investopedia* `https://www.investopedia.com/terms/t/transactor.asp/`
[5] CreditCards 2020 *Revolver definition - Glossary* `https://www.creditcards.com/credit-cardnews/glossary/term-revolver/article`
[6] Molnar C, Casalicchio G and Bischl B 2018 iml: An R package for interpretable machine learning *J. Open Source Softw.* **3** 786-787
[7] Stavins J 2020 Credit card debt and consumer payment choice: what can we learn from credit bureau data? *J Financ Serv Res* **58** 59-90
[8] Agarwal S, Chomsisengphet S, Mahoney N and Stroebel J 2017 Do banks pass through credit expansions to consumers who want to borrow? *Q. J. Econ* **133** 129–190
[9] Addo PM, Guegan D and Hassani B 2018 Credit risk analysis using machine and deep learning models *Risks* **6** 38-57
[10] Khandani AE, Kim AJ and Lo AW 2010 Consumer credit-risk models via machine-learning algorithms *Journal of Banking & Finance* **34** 2767–2787

IOP Conf. Series: Materials Science and Engineering 1049 (2021) 012021 doi:10.1088/1757-899X/1049/1/012021

Classification of lung infected Corona Virus X Ray Images using Deep Learning CNN Model

Yogesh Kumar Gupta[1], **Saroj Agrawal** [2], **Aman Mittal** [3]
[1]Department of Computer Science, Banasthali Vidyapith,Rajasthan, gyogesh@banasthali.in.
[2] Banasthali Vidyapith, Rajasthan. sarojagr708@gmail.com
[3]IIITKanpur,amanmitl@iit.ac.in

Abstract. All Start In this examination as learning, a dataset of COVID-19 lung contaminated X-beam pictures with common infection pneumonia, insisted Covid-19 sickness, and ordinary occurrences were utilized for the programmed acknowledgment of the Corona virus infirmity. The purpose of the investigation is to the comparable portrayal of Covid-19 utilizing Convolution Neural Network and assesses the precision proposed model convolutional neural framework over the progressing years for clinical picture characterization. Corona Virus lung infected X-Ray Images accessible by Kaggle a complete 590 images, which classified in 2 classes: typical and Covid-19.To survey the hypothesis accuracy of the models. The plan models end up being productive appeared differently in relation to made by (Arpan Mangal et al, 2020) , (A. A. Saraiva et al.,2020) , (Ioannis D. Apostolopoulos et al, 2020), (Ali Narin et al. 2020),which gained accuracy 90.5,95.30, 96.78, 98 independently and the current work had an ordinary precision of 99.54.

1. Introduction

As we know mostly all the lung diseases are infectious and can be transmitted from one person to another. Thus making Lung diseases very deadly as they can spread across the globe in very minimal time. In the Light of the recent event COVID19 is the most recent lung disease that we are encountered with and since then entire world is fighting against it day and night. Our lung plays a vital role in respiration process and its major role is to provide oxygen and remove carbon dioxide, with the increase in air pollution our respiratory system comes in contact with many hazardous elements which are not good and can cause major lung diseases. The virus is spread with people in close contact via droplets produced by sneezing, coughing, talking etc. The droplets remain on nearby surfaces thus the surfaces come in contact with other people and hence infecting them. That's why people are focusing on maintaining social distancing, wearing masks and sanitising hands regularly.

It is seen that COVID19 is more risky to persons having weak immunity particularly children and old-citizens. Same goes with the persons who are having diseases such as heart disease, diabetes, asthma etc.COVID19 is not only a danger for community health but it has also seen to generate discrimination in the infected countries. The fist case of COVID19 came in December 2019 in China (Wuhan city) and was observed that it came from a fish later it spread across the world.

In this paper, our essential centre is to distinguish COVID19 from lung x-beam pictures. With the fast multiply of COVID19 and the expanding request of conclusion analysts are headed to discover more clever determination measure. The classification consists of two parts, first part is the training and the second part is the validation and testing the model on the unknown images.

We divided this paper into 7 sections; section 2 is to show the previous studies in this field. In section 3 explained the steps of the algorithm. In section 5 and 6 presented model results and conclusion.

2. Literature Review

Many past researches are being bone on detecting COVID19 from radiological images. We have shown all such studies in the below table 1

Table1. Literature Review

Author Name	Paper Title	Algorithm	Observation	Accuracy
Ioannis D. Apostolopoulos1 · Tzani A. Mpesiana2	Covid-19: automatic detection from X-ray images utilizing transfer learning with convolutional neural networks	Deep learning with CNNs	The aim is to evaluate the performance of CNN architectures for medical image classification.	96.78%
Tulin Ozturk, Muhammed Talo, Eylul Azra Yildirim, Ulas Baran Baloglu, Ozal Yildirim, and U. Rajendra Acharya	Automated detection of COVID-19 cases using deep neural networks with X-ray images	The DarkNet model in deep neural networks	to detect and classify COVID-19 cases from Xray images.	98.08%
Ali Narin1, Ceren Kaya2, *, Ziynet Pamuk2	Automatic Detection of Coronavirus Disease (COVID-19) Using X-ray Images and Deep Convolutional Neural Networks	three different convolutional neural network based models (ResNet50, InceptionV3 and Inception-ResNetV2)	the detection of coronavirus pneumonia infected patient using chest X-ray radiographs.	98%
A. A. Saraiva, N. M. Fonseca Ferreira	Classification of Images of Childhood Pneumonia using Convolutional Neural Networks	CNN Model using K-fold	Classifyimages in normal and pneumonia.	95.30%
Nour Eldeen Mahmoud Khalifa et al	Detection of Coronavirus (COVID-19) Associated Pneumonia based on Generative Adversarial Networks and a Fine-Tuned Deep Transfer Learning Model using Chest X-ray Dataset	a pneumonia chest x-ray detection based on generative adversarial networks (GAN) with a fine-tuned deep transfer learning	categories: Normal and Pneumonia	99%
Halgurd S. Maghdid, Aras T. Asaad,	Diagnosing COVID-19 Pneumonia from X-Ray and CT Images using Deep Learning and Transfer Learning Algorithms	deep learning and transfer learning algorithms, CNN and modified pre- trained AlexNet model	COVID-19 Detection	98%
Srinivas Atreya	DISTINGUISHING COVID-19 USING CHEST X-RAYS	Uniform Manifold Approximation and Projection(UMAP)uses manifold learning techniques	separate out the Normal, Pneumonia and COVID-19 chest X-Ray images	Not Analyzed
Ho Yuen Frank Wong, Hiu Yin Sonia Lam, Ambrose Ho-Tung Fong	Frequency and Distribution of Chest RadiographicFindings in COVID-19 Positive Patients	RT- PCR Method	finding the severity of CXR	Not Analyzed
Sohaib Asif, Yi Wenhui	Classification of COVID-19 from Chest X-ray images using Deep Convolutional Neural Networks	Deep convolutional neural networks (DCNN).	detection of coronavirus pneumonia infected patients using chest X-ray rand gives a classification accuracy	98%

3. Approach to Our Model

3.1 Dataset Description

The dataset we utilized contains 590 X-beam (JPEG) pictures with two classifications: Normal and COVID-19. The X-beam pictures are taken from Kaggle.com. All pictures in dataset are physically examined so as to evacuate inferior quality pictures, just as being characterized by expert doctors to affirm the ground truth and miss-characterization.

- Training Dataset – My training dataset comprises of 370 images with 130 COVID X-Ray images and 240 Normal X-Ray images.
- Validation Dataset - My validation dataset com-prises of 220 images with 82 COVID X-beam pictures and 138 Normal X-beam images.
- Labels – COVID19 : 0 , Normal : 1

3.1.1. COVID-19 and Normal. Radiologists utilise a wide range of highlights in a x-beam picture to identify the COVID19. It is seen that ground glass design (GGO) is usually found in COVID19 patients. The GGO examples are distinctive in the later phases of the infection when contrasted with its beginning phases. In radiography, murkiness is utilised to speak to aspiratory combination which compares to the whitish territory, as appeared in the Figure 1 (b), the X-beam picture of a COVID19 tainted individual and Figure 2 (a) shows the x-beam of a typical patient.

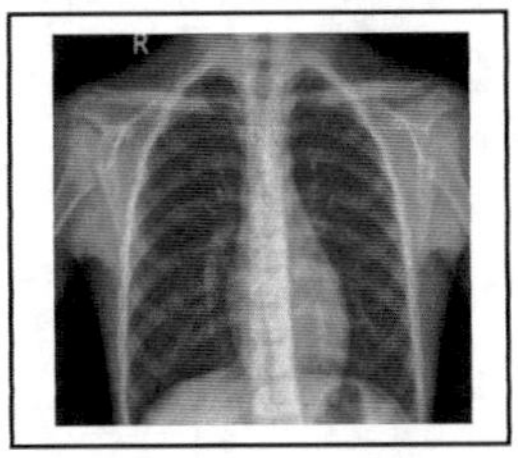

Figure 1. Normal Images

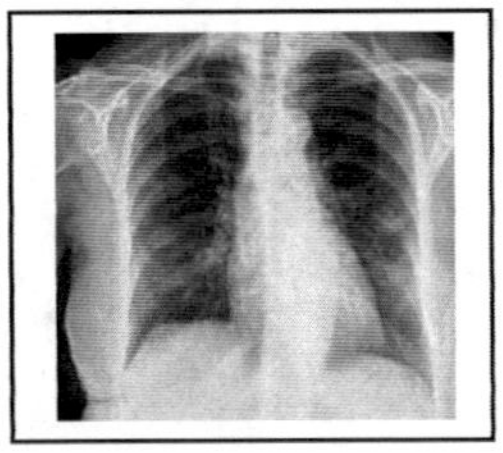

Figure 2. COVID Images

3.2 Input Structure of the Model

In Figure 3 essential design of our model is spoken to, which comprises of the image in the CNN network (which is utilized for categorization of pictures). Right off the bat all the pictures are resized to (224x224) continued by standardization with the goal that all the values lie between 0 to 1.

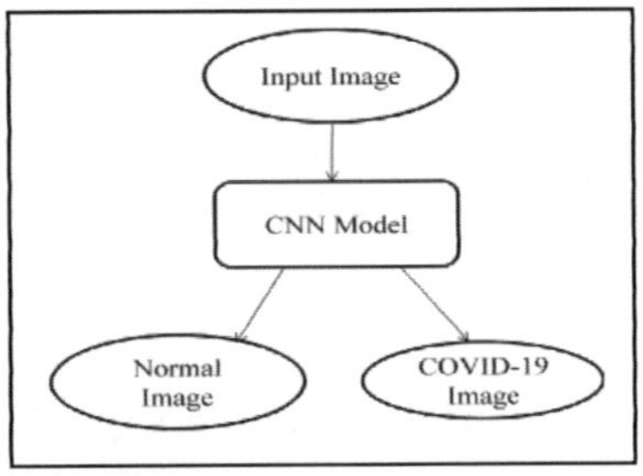

Figure 3. Input Structure

3.3 Training composition of the Model

Our training composition has many steps; the first step is the division of images into training and validation data sets than all the images are resized and normalized. The second step is the training of the images. Than the model validation is done through the validation data set.

3.4 Proposed Convolution Neural Network (CNN) Model

Every CNN has two parts first is the feature extraction part and the second one is the classification part. Feature Extraction part consists of convolutional layers and pooling layers while the categorization fully connected layers.

- The Convolutional Layer: This layer helps in extracting features from the input image which afterwards is helpful in making predictions. We give image matrix and a filter which is smaller than the image matrix as an input to the layer. Filters are a set of weights which are learnt by our model to detect specific features. The filter is moved on every part of the picture,

IOP Conf. Series: Materials Science and Engineering 1049 (2021) 012021 doi:10.1088/1757-899X/1049/1/012021

left to right to check a particular feature. We basically perform dot product between the small parts of input matrix and the filter. The learning of filter weights is done by training process.

- The Pooling Layer: It is utilized to reduce the parameters of the network and quantity of boundaries of our model in this way helps in diminishing the computational expense. This is generally put after every convolutional layer. For the simple understanding it just simply condenses the feature maps. Thus all the operations afterwards are applied on these condensed feature maps. There are mainly 3 types of pooling max-pooling, average pooling and global pooling. For our model we used max-pooling (most preferred type of pooling), in this it selects the max element from every small parts of the feature map which are covered by filter, as shown in figure 4.
- Fully Connected Layers: It is used for the classification, before giving input to the FC layers we convert our feature matrix into a vector using a flatten layer. Output of the FC layer is the 1-D matrix with the scores of all the classes in our model using a sigmoid activation function ranging from 0 to 1.

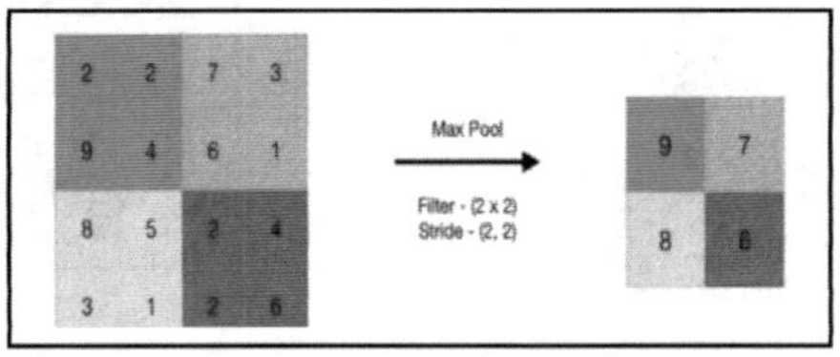

Figure 4. Max-Pooling Example

3.4.1. Binary Cross Entropy. The loss function we use is the cross-entropy loss function it helps in making a decision about the presentation of our model, with the yield being likelihood esteem between 0 to 1.

- *Loss = - (y log (^y) + (1-y) log(1- y)) y = original value, y^ = predicted value*

3.4.2 Convolutional Neural Network Architecture: In Figure 5 is the CNN architecture designed in 11 layers, seven feature extraction layers and four classi-cation layers, the network input receives a 224x224 pixel image after normalization, used RELU activa-tion function for each convolutional layer and a kernal of size 3x3, also gradually increased the amount of the features to be extracted by convolutional layers as go deep into the model. In this used max-pooling layer helps in dogging the problem of over fitting. To reduce the training time we also use dropout layers. At the end we used sigmoid function for the classification output.

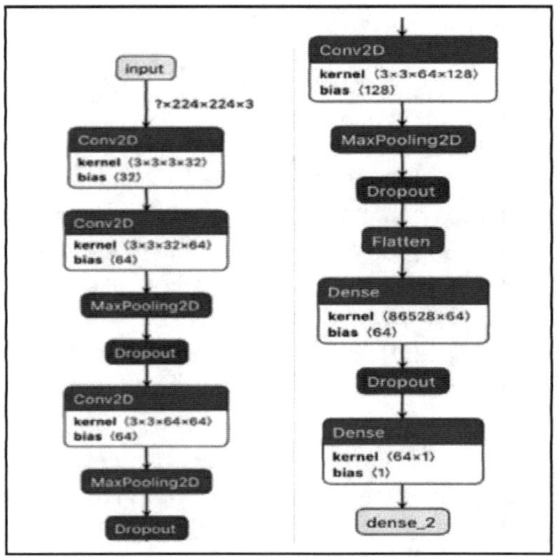

Figure 5. CNN Architecture

4. Measuring Parameters of the Evaluation

4.1 Confusion Matrix
The confusion matrix contains true positives, false positives, true negatives, false negatives.

- True +VE: Peoples are COVID-19 according our model.
- False -VE: Peoples are actually with COVID-19 but categorized as Normal our model.
- False Positive: People are actually Normal, but categorized as COVID-19, according our model.
- True -VE: Peoples are really Normal and categorized as Normal according our model.

4.2 ROC Curve
The (ROC) Receiver Operating Characteristic curve is plotted involving true positives and false positives by varying thresholds. It is used to visualise the performance for our classification problem. The AUC (area under the curve) score of ROC curve lies between 0 to 1. More the AUC score means more our model is accurate. It is basically a curve which shows relationship between test accuracy and the probability of miss-classification as shown in Figure 6.

4.3 Precision-Recall plot
It is a plot between precision and corresponding recall values as shown in figure 7. We can also calculate an AUC score for this plot. Similar to ROC more the AUC score of the p-r plot more accurate our model.

4.4 Classification report
Classification report consists of all the other features used to measure a model such as precision (for all the classes), recall (for all the classes)), F-1 score, accuracy.

Accuracy - Accuracy shows how correct are the predictions made by our model as compared to round truth.

Precision - Precision is a measure that mentions to us what extent of patients that we analyzed as having COVID19, really had Covid19. Precision = T.P. / (T.P. + F.P.)

Recall - Recall is a measure that mentions to us what extent of patients that really had COVID19 was analyzed by the calculation as having COVID19. Recall = T.P. / (T.P. + F.N.)

F-1 score - It is the harmonic mean of precision and recall, more F-1 score means more accuracy.

Table 2: Categorize Report

	Precision	Recall	F-1 score	Support
0	0.99	1.00	0.99	82
1	1.00	0.99	1.00	138
Accuracy			0.95	220
Macro Avg	0.99	1.00	1.00	220
Weighted Avg	1.00	1.00	1.00	220

5. Result
In this area will be introduced the outcomes got by our characterization model and the measurements we depicted previously. Figure 6, 7, respectively shows ROC curve, precision-recall plot. Table 2 shows the classification report of our model.

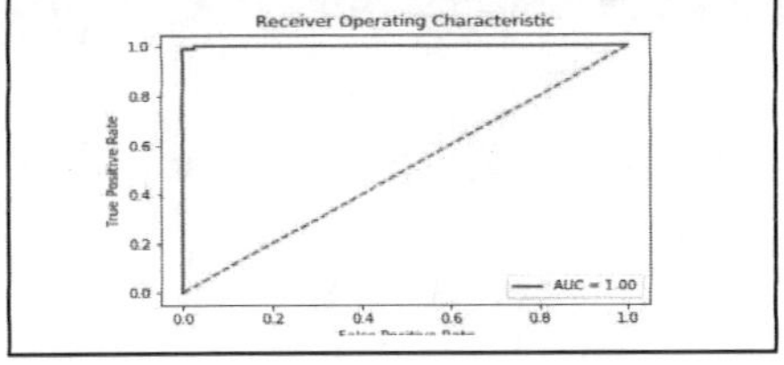

Figure 6. ROC curve for table 2.

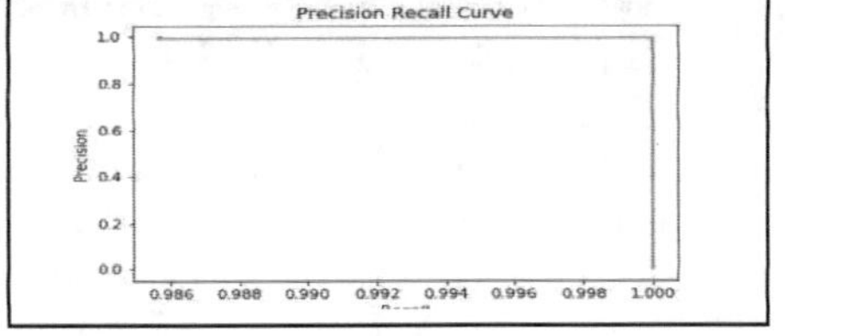

Figure 7. Precision Recall curve for table 2.

IOP Conf. Series: Materials Science and Engineering 1049 (2021) 012021 doi:10.1088/1757-899X/1049/1/012021

6. Conclusion

In this research Convolutional Neural Networks is utilised for building our arrangement model. We have additionally contrasted our model with the related past examinations (Shown under section 2 of Literature Review) and it comes out to be classification model has achieved a highest accuracy of 99.54 percent.

References

[1]. Ioannis D. Apostolopoulos1 ,Tzani A. Mpesiana, et al, 2020, "Covid-19: automatic detection from X-Ray images utilizing transfer learning with convolutional neural networks", *Physical and Engineering Sciences in Medicine* **43**:635–640.

[2]. Tulin Ozturk, Muhammed Talo, Eylul Azra Yildirim, Ulas Baran Baloglu, Ozal Yildirim, and U. Rajendra Acharya et.al. 2020, "Automated detection of COVID-19 cases using deep neural networks with X-Ray images", Comput Biol Med.; 121: 103792, doi: 10.1016/j.compbiomed.2020.103792.

[3]. Ali Narin1, Ceren Kaya, Ziynet Pamuk et al, 2020, "Automatic Detection of Coronavirus Disease (COVID-19) Using X-Ray Images and Deep Convolutional Neural Networks", arXiv preprint arXi..

[4]. A. A. Saraiva, N. M. Fonseca Ferreira et al, 2020, "Classification of Images of Childhood Pneumonia using Convolutional Neural Networks", Computer Science, DOI: 10.5220/0007404301120119 Corpus ID: 88488645.

[5]. Nour Eldeen Mahmoud Khalifa et al; 2020, "Detection of Coronavirus (COVID-19) Associated Pneumonia based on Generative Adversarial Networks and a Fine-Tuned Deep Transfer Learning Model using Chest X-Ray Dataset".

[6]. Halgurd S. Maghdid, Aras T. Asaad et al, 2020, "Diagnosing COVID-19 Pneumonia from X-Ray and CT Images using Deep Learning and Transfer Learning Algorithms", Deep AI: The front page of A.I.

[7]. Srinivas Atreya et al 2020, "DISTINGUISHING COVID-19 USING CHEST X-RAYS", https://www.analyticsinsight.net/distinguishing-covid-19-using-chest-X-Rays.

[8]. Ho Yuen Frank Wong, Hiu Yin Sonia Lam Ambrose Ho-Tung Fong et al, 2020, "Frequency and Distribution of Chest Radiographic Findings in COVID-19 Positive Patients"; https://doi.org/10.1148/radiol.2020201160.

[9]. Sohaib Asif, Yi Wenhui et al, 2020, "Classification of COVID-19 from Chest X-Ray images using Deep Convolutional Neural Networks", doi: https://doi.org/10.1101/2020.05.01.20088211.

[10]. F. Shan+, Y. Gao+, J. Wang, W. Shi, N. Shi, M. Han, Z. Xue, D. Shen, and Y. Shi, 2020, "Lung infection quantification of covid-19 in ct images with deep learning," arXiv preprint arXiv:2003.04655.

[11]. Zhang, Z. and Sabuncu, M. R., 2018, "Generalized cross entropy loss for training deep neural networks with noisy labels", CoRR, abs/1805.0783.

[12]. Brzezinski, D. and Stefanowski, J., 2017, "Prequential auc: properties of the area under the roc curve for data streams with concept drift. Knowledge and Information Systems", **52** (2):531–562.

[13]. Cherian, E.K Mulholland, J.B Carlin, H. Ostensen, R. Amin, M. De Campo, D. Greenberg, R. Lagos, "Lucero, M.; Madhi, S.A. et al.; (2005); "Standardized interpretation of paediatric chest radiographs for the diagnosis of pneumonia in epidemiological studies", *Bull. World Health Organ,* **83**, 353–359.

[14]. C Santosh, S. Antani; "Automated chest X-Ray screening: Can lung region symmetry help detect pulmonary abnormalities", *IEEE Transactions on Medical Imaging*, **37** (5):1168–1177.

[15]. A. Krizhevsky, Sutskever, G.E Hinton, 3–6 December 2012, "Imagenet classification with deep convolutional neural networks" , In Proceedings of the Advances in Neural Information Processing Systems (NIPS), Lake Tahoe, NV, USA,; pp. 1097–1105.

[16]. Zhu N, Zhang D, Wang W, Li X, Yang B, et al, 2020, "A novel coronavirus from patients with pneumonia in China ,2019 ", N Engl J Med 382:727-733.

AUTHOR INDEX